Body
Language

Body Language

How to know what's REALLY being said

2nd edition

JAMES BORG

Prentice Hall Life
is an imprint of

Harlow, England • London • New York • Boston • San Francisco • Toronto • Sydney • Singapore • Hong Kong
Tokyo • Seoul • Taipei • New Delhi • Cape Town • Madrid • Mexico City • Amsterdam • Munich • Paris • Milan

PEARSON EDUCATION LIMITED

Edinburgh Gate
Harlow CM20 2JE
Tel: +44 (0)1279 623623
Fax: +44 (0)1279 431059
Website: www.pearson.com/uk

First published in Great Britain in 2008
Second edition published 2011

© James Borg 2008, 2011

Pearson Education is not responsible for the content of third-party
internet sites.

ISBN: 978-0-273-75879-2

British Library Cataloguing-in-Publication Data
A catalogue record for this book is available from the British Library

Library of Congress Cataloging-in-Publication Data
A catalog record for this book is available from the Library of Congress

Illustrations by Adrian Cartwright (Planet illustration)
Typeset in 9.5/13pt IowanOldBT by 30
Printed and bound in Great Britain by Henry Ling Ltd, at the Dorset Press,
Dorchester, Dorset

Contents

Publisher's acknowledgements

We are grateful to the following for permission to reproduce copyright material:

Text

WITH ONE LOOK from SUNSET BOULEVARD, music by Andrew Lloyd Webber, lyrics by Don Black and Christopher Hampton with contributions by Amy Powers. Reproduced by permission of The Really Useful Group Ltd, London; extract from 'Cameron and Clegg: what is their body language really saying?', *Daily Telegraph*, James Borg, 15 May 2010, © Telegraph Media Group Limited 2010; extract from 'Tiger Woods put in a controlled performance, but he was clearly petrified', *Daily Telegraph*, James Borg, 19 February 2010, © Telegraph Media Group Limited 2010.

Picture credits

Alamy Images: Classic Image / Alamy, p.61, Allstar Picture Library / Alamy, p.175; **Corbis:** Boris Roessler / dpa / Corbis, p.63, ANDY RAIN / epa / Corbis, p.98, Victor Fraile / Corbis, p.114, Pool Photograph / Corbis p.133, Alessandra Benedetti / Corbis, p.219; **Getty Images:** John Kobal Foundation, p.xx, Al Seib / LA Times Oscars / Contour, p.57, James Devaney, p.124, Bloomberg, p.184, WILLIAM WEST, p.197, CARL DE SOUZA, p.198, MGM Studios, p.229; **Press Association (PA Photos):** Eamonn and James Clarke / EMPICS Entertainment, p.66, ellis obrien / allaction.co.uk / EMPICS Entertainment, p.83, RICHARD DREW / AP / Press Association Images, p.194; **Rex Features:** Everett Collection / Rex Features, p.44, David Fisher / Rex Features, p.135, SNAP / Rex Features, p.224.

'From the moment I picked up your book until I laid it down, I was convulsed with laughter. Some day I intend reading it.'

Groucho Marx

Preface to 2nd edition

Well here it is: the expanded and updated edition which I hope may enlighten you and raise your awareness of this fascinating topic. As with the other two books that form the 'trilogy', the aim is to inform, educate and entertain.

Communication is such an important topic that is essential to our well-being; and most of it is not from what we say. *How* we say it as well as the *postures* and *gestures* we use is the 'magic' solution.

Even in the last few years there has been more focus on observing the body language of politicians, 'celebrities', sports-people and people in the workplace. On the flip side there is the growing realisation that our own body language has to send the right message too; if we're to communicate our intentions accurately.

I was 'humbled' to receive an award for this book: Winner of the BAA (British Airports Authority) 'BEST NON-FICTION TRAVEL READ AWARD 2009'.

From a short-list of six books (the odds were slightly lessened by the inclusion of another book of mine, *Persuasion*, within the six!), more than 23,000 members of the public cast their votes in a month-long process.

If any of you pick up this new edition, I would like to express my thanks.

Once again – enjoy the 'journey'.

JB

Author's note

It's been famously said that when writing a book 'you don't choose your subject; your subject chooses you'. Certainly that was the case with my previous books that form the 'trilogy': *Persuasion* and *Mind Power*. I've spent a lifetime studying persuasion, how the mind works and body language: both academically and in my working life where *observation skills* and *self-awareness* were (and still are) crucial to achieving positive outcomes and an understanding of other people.

Interest in non-verbal behaviour, or 'body language', has grown rapidly in recent years because in this fast-paced and time-poor world we're constantly judged on *first impressions*. **People are making snap decisions as to whether they trust us, like us, want to work with us, have a love affair with us – and much more.** As research continually points out, words alone don't provide the whole picture. It is in the nature of the human condition that we communicate more through the body than merely through language.

Every day we constantly have to *interpret* what another person's body language is telling us – as well as *controlling* our own to create the right impression. Two-way traffic!

You may have come across the oft-cited study in the 1970s which concluded that more than 90 per cent of meaning in any interaction is derived from non-verbal cues – the manner in which your body 'talks' and also the way that you *say* things (vocal) – and just a mere 7 per cent from the *words* you actually speak.

We can play around with the actual figures of course, but the fundamental point is that the overwhelming meaning of a message, when communicating with others, comes from an unconscious display of the 'silent' language; this either reinforces or detracts from the words being used. Research has shown that the most successful people, *in all walks of life,* are 'intuitive' in deciphering these signals.

'All the world's a stage'

Actors have to be masters of body language in order to convince us to believe in the role they are playing while at the same time helping us to 'suspend our disbelief' (since it's not real life) so that we can engage with and put our *own* emotions into the performance. When we see mannerisms and gestures that ring true to a particular emotion, or what's being said or felt, we unconsciously accept the 'message' and go along with the make-believe – good acting!

Just recently I noted down an excerpt from a review of a play in London's West End:

> . . . I believed in her persona. The body language and demeanour and status were all well observed . . . she combines dance and theatre to convey emotions and thoughts so that the audience is able to interpret without the need of words.

Since we're all acting out certain 'roles' in everyday life, both in our personal lives and especially at work, body language is the way in which our bodies communicate our own or a 'character's' attitudes. There's nothing 'false' about it – as Shakespeare told us:

> All the world's a stage,
> And all the men and women merely players;
> They have their exits and their entrances;
> And one man in his time plays many parts.
>
> *(As You Like It)*

The 'magic' of body-reading

Just like a professional actor, when you're acting out your role in everyday life you have to make sure your body language is **appropriate** for the character you're playing, otherwise your **'performance'** is not congruent and your message not believable.

Certainly, this was brought home to me as a youngster, fascinated by magic and the psychology behind its presentation. After painstakingly acquiring magical knowledge and then being let loose to perform 'effects' (conventional and 'mind-reading') I was eventually accepted as one of the youngest ever members of the Magic Circle. There the magicians' dictum, courtesy of the famous Robert Houdin (from whom the great Houdini had taken his name) was drilled into me: 'A magician is an *actor*, playing the *part* of a magician.'

What was that all about? Well, I was told that, **since most communication is conveyed through your demeanour – in posture, gestures, eye contact, voice, confidence – just like any other 'actor' in the performing arts, you had to become an expert in conveying the right impression – *to be 'believable'*.** Your body language had to reinforce the part that you were playing. That separated the good performer from the mediocre – presentation was all.

Even more to the point – just to complicate things for this youngster – the discipline of magic that was of interest to me was not conventional magic but 'mind-magic' or **mind-reading. This meant that 'tuning in' to other people's thoughts and being able to 'read' body language (*coupled with certain 'magical' techniques of course!*) was essential in order to perform the *miracle*.** So there was a double necessity for me in acquiring body language knowledge and skills. I'd chosen that branch of magic – mind-reading and 'mind-magic' – that relied, in part, on being able to read body language well.

What people do with their bodies is a window to their subconscious thoughts, so close observation is the key to reading minds. Thus began a lifelong journey in honing perception skills and becoming more self-aware. My own body language had to be right too, in order for an audience to suspend their disbelief and accept that their minds (or thoughts) were being 'read'.

All this brought home to me two important things:

1 In real life we're constantly trying to *read minds* by observing body language. We're all engaging in mind-reading in one way or another.

2 Body language is a *two-way street*. You need to be aware of:

- your *own* body language – and the messages you are giving out (after all other people are 'reading' you)

- how to 'read' the body language of *others* to determine the messages they are giving out.

Years of performing 'mind-reading acts' honed and furthered my interest in body language. Academic study in psychology and related fields meant that when I was let loose in the world of work, the awareness, thankfully, was already there.

Becoming 'fluent' in body language

After we've been through the 7 Lessons you'll have all the tools you need to become an accomplished reader and user of body language. Our aim is therefore twofold:

1 To develop the self-awareness you need to **control** your own body language so that it delivers the **right** outcome for you.

2 To sharpen your senses so that you can **read** body language in others and **react** in the appropriate way.

After all, if you're trying to get inside the mind of the other person – by observing what's happening on the outside – *they'll be doing the same to you.* So you'll need to use the right body language to convey the impression that you mean to make – as opposed to leaving it entirely to your subconscious, *as you've done in the past.*

You'll become an adept mind-reader and have greater success in fathoming what a person's really thinking. All you'll need is to be able to decipher gestures that you previously **paid no attention to** (both the subtle and the blindingly obvious), and also to be aware of your **own** gestures and how they might be *provoking* a reciprocal gesture in the other person.

I've cut out a lot of the 'peripheral' information that scientists have discovered – sometimes it's not worth 'dissecting' things for the sake of it – to concentrate on what is *practical*. After you've absorbed the 7 Lessons and then combined them with daily practice you'll be transformed into a body language 'wiz'.

In conclusion

At the end of the 7 Lessons you should find you have raised your awareness to the extent that you:

- **become more intuitive in deciphering *other* people's body signals**
- **are aware of your *own* 'bodytalk'.**

Also you will be able to *control* it and *use* it to great advantage to enhance your own communication style – with friends, strangers, family, work colleagues, customers, clients.

This is a book for everybody – whatever you do, if you have contact with other people (few of us can be excluded from this category) and you want to know how to read people better and simultaneously make yourself more effective in conveying the impression you intended to create, then read on.

So here it is. This is for you: the person *in a hurry* – enjoy the journey!

James Borg

'The average person looks without seeing, listens without hearing ... touches without feeling ... moves without physical awareness ... and talks without thinking'

Leonardo da Vinci

Introduction

If you could read my mind...the 7 Ls

There can be few things more fascinating to men and women than the language displayed by other people's bodies – as well as their own. What we'll do over the course of the 7 Lessons is provide you with enough knowledge to confidently read the body language of others and – crucially – to be aware of your own. We'll train you to *look* and *listen*. You'll find that your new-found powers of observation will change your life as you learn to really *look* and really *listen*.

Just remember one thing at the outset – the science of body language is not an **exact** science. Whenever you're dealing with complicated 'systems', such as human beings, nothing can ever be straightforward. That's why, as we'll see, it's essential to piece together a number of behaviours in order to make an accurate reading. Otherwise you'll fall victim to **ID 10T errors** all the time (more about that later).

First of all, go back and take a look at the da Vinci quote on page xiv. You may recognise yourself – do you go about your life with your senses 'dulled'?

It's not easy to fake body language. The human body is comprised of many muscles and to be aware of the activity of all of them at the same time is impossible – and we're including facial muscles in this too. No matter how good you think you are at controlling your anatomy, there will always be *'leakage'* (involuntary signals) that give away your true feelings.

❝ It's not easy to fake body language ❞

So let's just summarise the two-way street of body language and *why it's so important*:

- If you're trying to communicate a point, choose the appropriate body language to have a far better chance of achieving the outcome you're after.
- Being able to read the body language (or non-verbals) of others allows you to modify and shape your message based on your receiving of subtle positive or negative signals during your interaction.

BODY WISE

Feelings are communicated more by non-verbals than by a person's words.

Body language will always be the most trusted indicator for conveying:

- feelings
- attitudes
- emotions.

We unwittingly go about our everyday lives displaying our inner thoughts. The relatively new form of communication, speech, fulfils the role of conveying **information** (facts and data) while the body fulfils the role of **feelings**.

It's an inescapable fact that our non-verbal actions scream out more about our moods and feelings than we would perhaps wish to disclose. People tend to use the whole body to read a person's moods and attitudes and this is absorbed mainly at the subconscious level.

That's why there is the need for '**congruence**' (one of our 3 Cs – more on these later) if we're to believe a message that's being conveyed to us. In many cases what we're perhaps displaying may be an unintended mannerism that devalues the words that are spoken and creates a mixed message for the listener.

So when we speak and accompany it with body language that casts **doubt** on the truth of the message, so that the words are not 'congruent' with the language of the body, it causes doubt on the part of the listener. It could just be a bad habit which conveys the wrong impression and has not been corrected. **Intermittent pursing of the lips, holding the head in the hands, covering the mouth with fingers while speaking, sighing at inopportune moments, constantly shifting in the chair while talking.** These may all just be bad habits rather than gestures denoting a specific feeling relating to the message. But the point is that they can be *misunderstood*.

It's bad enough being found out by your body language when you're trying not to give things away. But when you're 'not guilty' – and it's just an irritating habit or mannerism that's distorting your message – that clearly is not good.

If the person doesn't know you that well or is meeting you for the first time, they have no 'baseline' behaviour knowledge about you (so they won't know that a particular gesture is a natural part of your demeanour). **All they can go on is what they see or hear.**

First impressions are powerful – and difficult to change.

BODY WISE

People who know you the least will judge you the most.

When we're communicating with friends, relatives, work colleagues or strangers we all have certain habits that are a part of us in a specific context or situation. If you're more aware of body language then you'll know which of these habits to change in order to improve relationships. It may not happen overnight, but you can gradually supplant these gestures or mannerisms with ones that don't impede your message, with a bit of patience. As the famous writer and sage Mark Twain once said:

Habits can not be thrown out the upstairs window. They have to be coaxed down the stairs one step at a time.

When did it start?

We've only been studying non-verbal communication, or body language, for around 50 years or so – although social anthropologists will remind us that its origin goes back to the beginning of time, before the spoken word. Even Leonardo da Vinci in the 16th century, despite his other considerable talents, was interested in developing his 'interpersonal intelligence' and improving his senses. He advised:

When you are out for a walk, see to it that you watch and consider other men's postures and actions as they talk, argue, laugh or scuffle; their own actions and those of their supporters and onlookers.

Leonardo da Vinci

For most of us, the fascination of black and white 'silent' movies is as near as we get to appreciating how 'actions speak louder than words'. If you've seen the stars of the silent movie era, you'll appreciate the power of this silent language.

Groucho Marx

Who can forget images of Charlie Chaplin and even a bit later, when the 'talkies' started, the films of the Marx Brothers, which had sound but conveyed much of the humour through the actors' gestures (remember Groucho's dancing eyebrows punctuating his wisecracks)? If you've seen any of these you'll appreciate the power of the silent language. Body language gestures and expressions silently communicate feelings and emotions that *transmit a thought*.

❝ Who can forget images of Charlie Chaplin ... and Groucho's (Marx) eyebrows? ❞

The point about body language is that although we are perfectly able to select appropriate gestures and actions to transmit a message, our body also sends out signals outside our conscious awareness – in other words, *without our permission!* Whatever words we are using during any interaction with people, they are, whether we like it or not, always accompanied by bodytalk that can reveal much *more* than the spoken word. Yet, most people go about their business in all the activities of their daily life wholly unaware that they are **receivers** *and* **senders** of non-verbal language.

Small wonder, if you consider that around 95 per cent of the information that the brain takes in is through the eyes, relegating the other senses – which obviously are no less important – of sound, touch, taste and smell to just 5 per cent to complete the picture.

BODY WISE

Initially, we are more likely to believe what we **see** rather than what we **hear**. This will be perceived as the true meaning and, because of the way that the brain stores **memories**, this will be the impression that is **remembered**.

Recognise yourself?

It's a simple fact of life – people go around *attracting* others to them, or *repelling* them, because of their body language. Have you ever stopped to consider what your body language says when you are communicating with others?

Self-analysis

- Do you find that you're subconsciously turning people off?
- Are you – again subconsciously – giving off signals that say you're untrustworthy?
- Do you find it difficult to persuade people to change an attitude or behaviour?
- Do you have difficulty in securing a job offer after an interview?
- Do you have trouble in getting a date?
- Do you feel that you say the right things at the right time, in most situations, but still make no headway?

The list is endless.

The point is that if you don't have good body language (either through lack of **self-awareness** or **laziness**) and are not good at **reading** it in others, then you go about your daily life with everything becoming that much more difficult. Because it's such an essential part of the way we communicate, it means you're not bolstering your conversations and messages with appropriate feeling. Equally, it means you're not aware of the *clues* that are being given out by others.

> Be aware that everything we discuss in our 7 Lessons applies at two levels: (1) An analysis of your own body language (what signals do I *send* out?) (2) On the flip side, interpreting the signals that you *receive* from others.

Emotions and feelings

Recognising a person's emotions or feelings is your key to people-reading. **Emotions are conveyed more clearly through body language than through speech.** You're probably familiar with the term **'emotional intelligence'** which, over a decade ago, seemed to have kickstarted an awareness within people of the importance of emotions and feelings in human relationships. Five emotional competencies or skills have been recognised and the message is this:

1 Be aware of your own emotions.
2 Learn to control your emotions.
3 Assess the emotions of others.
4 Look for clues from the body language.
5 Relate successfully with other people.

“ *Five emotional competencies or skills have been recognised* **”**

The important point about these competencies is that 5 will only happen if you have successfully integrated 1 to 4.

Start with yourself

You probably are aware of these things subliminally, but in the cut and thrust of everyday life it's easy to take short cuts and, either through laziness, impatience or poor mood, ignore the signals *you* may give out and, at the same time, fail to engage your perceptivity and truly listen to others.

Before you interact with others you should take a moment to analyse your *own* emotional state. What is it? Impatient, angry, anxious, resentful? Each of these, for example, will influence the way that you address other people and body language 'leakage' will arise and may cause problems. So you need to manage or control these signals.

What about the others – what's **their** body language telling you about how they are feeling? What you pick up – for example, disinterest, frustration, anger, anxiety – may be nothing to do with you. Perhaps the other person has just heard that the insurance company is not paying out for the roof damage in the recent storm. The point is that it's up to you as the 'receiver' to try to engage their interest.

So *empathy* is needed first, to pick up a feeling and understand another person's perspective, and then *sensitivity* is needed to get people to 'open up'.

Sometimes people's body language is **open** and positive and at some point in a conversation, meeting or presentation will turn to a more '**closed**' position, perhaps with folded arms or hand-to-face gestures (more about that later) that indicate a problem. Again, you need to have the *perception* to notice this change and at what point it occurred in order to 'backtrack' and address the cause of this mood change.

ESP

Let's begin with a memorable phrase – **body language is the window to a person's mind**. Of course we'd all like to be able to read minds and **that's what you'll be learning how to do**.

The subject of extra sensory perception (ESP) always arouses strong emotions. But you'll be using a different version – your natural ESP – in order to rouse this dormant 'sense' of yours. Let's look at the three dimensions of your natural ESP – **E**mpathy, **S**ensitivity and **P**erceptivity – in more detail.

Empathy

This has finally been given its rightful place in terms of its importance in establishing rapport and trust. The concept of 'emotional intelligence' has highlighted the importance of empathy, which has been likened to our 'social radar'. It's been described as *sensing what others feel without them saying so.* But it has to be sincere. Since people will rarely disclose how they feel by just using words, we pick up their true feelings in three main ways, by:

- gestures
- facial expressions
- vocal clues.

These tell the real story about a person's feelings and perspectives. This, of course, is the essence of body language interpretation.

“ *These tell the real story about a person's feelings* **”**

Sensitivity

Being sensitive to the clues that are picked up through being empathetic, and acting accordingly after tuning in to another's thoughts, is the next stage. Being sensitive to one's own emotions is also important because body language is a two-way street. What signals are we **giving out** (through the emotions we are feeling), which in turn affects the **receiver's** behaviour – and the signals that they then give out to us? (Are we part of the solution, or *part of the problem?*) Having sensitivity requires us to have the capacity for self-awareness.

Perceptivity

All the information we've translated gives us a heightened sense of perceptivity to the other person's state and their emotions – it results in us having '**intuition**'. We subconsciously 'process' a person's words in the way they were said and with the body language we saw. We are then able to reflect back our perceptions with a much greater skill, which should help towards more positive outcomes.

So you can see that this combination of empathy, sensitivity and perceptivity gives you greater *insight* into the true feelings of others. This is the basis of what we normally refer to as our 'intuition' – it's a form of mind-reading.

Mind-reading or 'thought-reading'?

So you can see that observing or reading body language (coupled with our natural intuition) is the way that we try to engage in this process of mind-reading. But in order for you to be convinced that you're perfectly able to do this, it needs a slight amendment – look upon it as reading *thoughts*.

So your ESP skills are truly connected with mind-reading (or thought-reading, if you prefer). Here's the proof that I hope will truly convince you that at present you exercise this power from day to day – the aim is for you to be even better:

1 The mind produces a *thought*.
2 The thought produces a *feeling*.
3 That feeling 'leaks' out through *body language*.
4 You read the body language (to ascertain a person's feeling).
5 Hey presto, *you're mind-reading*.

Body language is a window to the mind

We subconsciously use our intuition to pick up signals from another person's posture, facial expression, gestures, tone of voice, eye movements and much more. And since other people are doing that to us too, we need self-awareness and empathy to become expert in non-verbal behaviour.

Above all we need to be aware of when the behaviour occurs, whether it seems at odds with other behaviour displayed and, if so, can we see multiple 'cues' to support it? We'll discuss this next.

The 3 Cs

It's vital to pay attention to the 3 Cs. No true reading can ever come about without taking into account **Context, Congruence** and **Clusters**.

❝ *It's vital to pay attention to the 3 Cs* ❞

- **Context**. It may seem obvious but you have to look at the context in which behaviour occurs. A man returns from an early morning run and walks with his head down and therefore has downcast eyes – he is also breathing heavily. Does that indicate boredom, insecurity or depression? No – he's just come back from a run. That's what it indicates.

- **Congruence**. Since visual and vocal body language (non-verbals) make up more than 90 per cent of a message, we need to see that the words match the actions – that they are congruent. For example, crossed arms, repeated looking away and sighing would not be congruent if a woman was telling others that she was enjoying a play at the theatre. We would believe the *visual* message.

- **Clusters**. Because it is obviously unwise to judge a *single* gesture for meaning, we have to look for clusters of gestures to interpret body language. One single gesture can be likened to a word in a sentence. The sentence gives us meaning (i.e. a number of gestures together). Make sure you always look for gesture clusters.

ID 10T error

Oh . . . and please take heed of Body Language Rule Number 3 – **Warning: ID 10T error**.

The two-way process of body language is one of:

1 **transmission** (by one person) and then
2 **interpretation** (by the other person).

So:

- A reads the body language 'message' of B (transmission and interpretation).
- A responds with his own body language signals to B (all well and good).
- B reads A's signals and responds with her own (all well and good).
- A **misinterprets** the signals from B.

The result – the breakdown in communication means that apart from *transmission* and *interpretation* there is a *third* element prevalent in non-verbal communication – **misinterpretation**.

So, as you'll see throughout the book, the reading of this silent language requires us to look for *clusters* of information (as opposed to one cue) to support our interpretation. Failure to do this will result in ID 10T errors.

This will become apparent to you as you progress through the stages.

Your natural 'intuition'

We're constantly being told that *knowledge is power*. This is never truer than when it applies to self-knowledge. The more you know about yourself, the more you will have the power to control your own thoughts as well as reading others.

It's never too late to learn about body language and it's a skill you can easily become adept at – if you train yourself to notice more while exercising a bit of caution and taking note of the 3 Cs.

We're all naturally good – potentially – at reading the silent language. After all, that's all there was at the beginning of time. It's just that many people have never bothered to take it a stage further and become aware of the need to be more observant. Without beating about the bush – it's usually *laziness*. But by just changing your 'habits of a lifetime' and becoming more *aware* of other people's actions – as well as your own – you'll notice a huge difference.

- You'll find that you have a heightened sense of intuition because you're paying more attention to what you see and also to how people are saying things.

- You'll pick up if a person is in a troubled state and also whether they're telling the truth, a polite lie or a more serious one.

- Your sharpened senses and powers of perception will allow you to tune in to other people's thoughts.

Thankfully, because you were born with that wonderful capacity for 'intuition', you can already tell if someone is giving out signals that spell out to you happy, miserable, anxious or relaxed. From a distance you can deduce whether people are having an argument, a friendly conversation or are in the throes of a mad, passionate love affair, just from observing **posture**, **gestures** and **facial expressions**.

You pick up all this information subconsciously. What would happen if you made a *conscious* decision to observe people more carefully? How proficient a reader of body language or of people's minds would you be then? You just need to know what to look for. So let's take it a stage further over the course of these 7 Lessons.

Lesson

1

'I speak two languages. Body and English.'

Anonymous

Language of the mind and body

No doubt over the years you've picked up on the various ways we all communicate – facial expressions, the way we stand or sit, gestures such as crossing our arms in a particular way, the position or tilt of the head or direction of the eyes. All these movements express something – even without accompanying speech. How you perform these movements contributes to creating the image that you present and will determine people's perceptions of you.

It's very rare for any of us, when we're talking to people, to believe that our words alone can convey the right message.

We may smile or grimace, avert our gaze at times, stand close or at a distance, touch (or not) and use other non-verbal communication to add weight to our message. A number of surveys over the past 50 years have provided a body of evidence (did I say 'body'? – it's crept in again) to show that it is body language – or non-verbal messages – that powerfully communicate the following:

- **acceptance** and **rejection**
- **liking** and **disliking**
- **interest** and **boredom**
- **truth** and **deception**.

Wouldn't it be good to be able to identify all of these in your interactions with other people? It would certainly save a lot of time and heartache, and also provide you with feedback that might enable you to salvage a situation in some cases.

So, a good awareness of body language provides practical insights into improving your interactions with other people in most situations. Friends, family, work colleagues, customers and clients, at job interviews, with strangers – it's an endless list.

Communicating with the silent language

Quite naturally, because of our daily interactions with people involving the spoken word, we've been educated to believe that language skills – or, more precisely, words – are of paramount importance. They are important, but the 'silent' language is of equal importance – if not more.

BODY WISE

We all pick up the subtle clues that others are sending out – even though we may not be conscious of it. And of course other people are doing this with us. We're sending cues that indicate either 'keep away from me' or 'I'm an approachable sort of person'.

&& *The 'silent' language is of equal importance* **&&**

We communicate with our:

- dress
- posture
- facial expression
- eye contact
- hand, arm and leg movements
- body tension
- spatial distance
- touch
- voice (tone, pace and inflection).

Because we communicate in this 'silent' language from the subconscious, it follows that as a true indicator of our feelings it conveys **more** than the spoken word.

Gestures are very effective in delivering messages in the form of images *in a way that speech is unable to do*. So it follows that when gestures and words are used simultaneously this is the most effective method of communication. We choose gestures to communicate our message but our body throws out signals that are beyond our conscious awareness (*and that's where the trouble starts*).

It's time to haul out the statistics relating to the groundbreaking – and still highly influential – study conducted in 1971 by social psychologist Professor Albert Mehrabian of the University of Los Angeles (UCLA). He looked at the relative strengths of verbal and non-verbal messages in face-to-face encounters and devised a communication model **which has stood the test of time**. It has come to be regarded almost as a template for understanding how people derive *meaning* from another person's message.

However, it has come to be *misinterpreted* over time as the topic of body language has gradually become more and more popular and extends its reach into the world of 'celebrity' and consumer magazines.

Nonetheless, the broad percentages have been confirmed through subsequent research over the last few decades. What is indisputable is that 'looking and listening' (to the non-verbals) – as mentioned earlier – is the key to deciphering the true meaning in any face-to-face communication with another person.

Mehrabian's research revealed three elements in any communication message – **body language**, **voice** and **words**. He came up with the famous '**55, 38** and **7**' model, which reveals that:

- 55 per cent of the meaning in any message comes from the **visual** body language (gestures, posture, facial expressions).

- 38 per cent of the meaning is **vocal,** derived from the non-verbal element of speech – in other words the way in which the words are delivered – tone, pitch, pace.
- 7 per cent of the meaning comes from the actual **words** (content).

This leads to a startling conclusion:

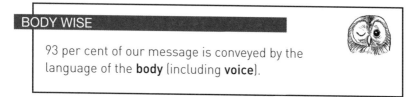

BODY WISE

93 per cent of our message is conveyed by the language of the **body** (including **voice**).

- This means that in those vital 20 seconds to 3 minutes that we have when people form a first impression, this is determined mainly from how we *present* ourselves and how we *say* things rather than *what* we say.
- If there is a *mismatch* between the words and the way they are delivered, we tend to believe the *delivery* rather than the words (i.e. the *highest* figure in the list above).
- Therefore, body language enables us to look *beyond* the words that are being used and get to the hidden **silent** message that is being conveyed (often through the subconscious).

So Mehrabian's classic research tells us that your impact boils down to three factors:

1 **HOW YOU** *LOOK*
2 **HOW YOU** *SOUND*
3 **WHAT YOU** *SAY.*

In short – **body language speaks louder than words!**

WARNING

Over the decades, some people, after learning about the 55, 38, 7 study (and other more recent studies that broadly confirm these figures), have *misinterpreted* these groundbreaking findings. They've concluded that words are not that important and that as long as you look confident, project the right impression, dress to kill and then deliver your *ill-construed* words with the right seductive pitch and tone of voice then the world's your oyster. (A classic illustration of the maxim that 'no information is better than *misinformation*'.)

So they've concluded that if words are worth less than 10 per cent in your interactions in terms of successfully engaging with people then actual words are not that important.

Incorrect. That's not what the study showed. When you read this in the future – and you're bound to come across it in a magazine in some guise or another – take a deep breath.

What did the research reveal? Well – essentially it was this:

> If your 55 per cent – visual body language – is not good,
> *they're not even going to stick around* (excuse the vernacular)
> to listen to the 45 per cent!

Even if your audience 'sticks around', if your 38 per cent (the way you speak) turns them off, they won't take in or comprehend the 7 per cent (the actual words) and they're off – **mentally if not physically.**

That's what the research findings showed.

How many times have you thought (or said) – at a party, at work, on a date – that things were looking good '*until he opened his mouth*'. (Have you ever seen any of those TV programmes on 'speed dating'?).

Make no mistake – the words are important. Our objective is to get the person to *want* to listen to us in the first place. Even if you consider yourself to be oozing with charisma without saying a word – words are still important. But of course **how** you say them is equally as important.

❝ *Get the person to* want *to listen in the first place* **❞**

It's generally accepted by most researchers that:

- words are used to communicate *information*
- body language (or non-verbals) conveys *attitudes, feelings* and *emotions.*

(We'll discuss the way words are voiced along with how attitudes, feelings and emotions are displayed, in Lesson 3 when we look at the non-verbal aspect of speech – **paralanguage**.)

Sometimes body language is used as an alternative medium for verbal messages (think about Norma Desmond's 'With one look I put words to shame. . .' from *Sunset Boulevard*).

So, on the basis of how you score in the three factors listed above, people will make decisions as to:

- whether they like you or not
- whether they trust you or not
- whether to go on a date with you or not
- whether to do 'business' with you or not.

Bluntly – **whether to have anything to do with you at all!**

CAUTION

Many people spend time trying to become expert in decoding the body language of other people, and they still don't improve their own personal and work relationships. Why? Because they forget to look at their *own* body language.

A lot of 'relationships' are formed or 'dissolved' **in the first three minutes of an encounter.** It's the 'gut instinct' or intuition from the subconscious that is picking up on the non-verbals to decide whether it's a thumbs-up or thumbs-down.

Your amicable words mean nothing if your body seems to be saying something different. We're constantly making impressions (as sender) for other people to receive as well as receiving impressions (as receiver) about other people. It's two-way traffic.

We'll evaluate, through our 'sixth sense', how we feel about a person by the way they express themselves through their body. It's not even a *rational* decision on our part. Call it intuition. This quote sums it up beautifully:

> **There is a road from the eye to the heart**
> **that does not go through the intellect.**
>
> **GK Chesterton**

Quit the navel-gazing as to *why* you do something with your body (facial expression, eyes, gesture). Think – from a visual viewpoint – *how* it appears to *other* people and, more importantly: *is that what I wanted to convey?*

The first impression sticks – for better or worse. You may remember the shampoo commercial on television many years ago – *'You never get a second chance to make a first impression.'* Never was there a truer maxim.

❝ *The first impression sticks – for better or worse* ❞

BODY WISE

As with toothpaste, it's easier to let negative first impressions out of the tube than to squeeze them back in.

So make sure that people are reading you correctly. If you look the part and your non-verbal display is consistent, your words will be reinforced and your 'audience' will have confidence and trust in what you're saying and will want to hear more.

Kinesics

Way back in 1872 Charles Darwin, known for his theory of evolution, wrote his groundbreaking *The Expression of the Emotions in Man and Animals*. It wasn't until the middle half of the following century that further serious scientific research started again.

One of the early pioneers of body language was Ray Birdwhistell, an American anthropologist who worked in the 1950s. He called this silent communication 'kinesics' because of its study of the way that various body parts or the entire body play a key role in communicating a message.

Our 'gestures', which broadly include movement, postures and expressions, transmit messages while the mouth is busily sending out the carefully crafted (or otherwise) words. The other 'giant' of the body language movement, zoologist Dr Desmond Morris, has defined a gesture as *'any action that sends a visual signal to an onlooker. . . and communicates some piece information to them'*. This can be either deliberate or incidental. Many of our incidental gestures are ones that we would prefer to conceal. For example, the head-on-hands during a less than exciting training session or the second half of a dull play. Quite often we may not be consciously aware of adopting a gesture (it is not deliberate), but this indicator of mood information sends out a signal to the onlooker, and the meaning of it is read.

The kinesic model was further developed by Paul Ekman and Wallace Friesen (University of California, 1970) – we shall be referring to some of their research on facial expressions in Lesson 2 – who subdivided kinesics into five broad areas that provide a convenient shorthand for us.

1 Illustrators

These tend to be gestures that **accompany** speech to create a visual supporting message that describes or reinforces your message – and more often than not are subconscious in their origin. For example, you might gesture with a rising upward movement of your upturned palm as you describe how house prices have gone up in the past two decades.

2 Emblems

These usually **replace** words – an obvious one is the thumbs-up. In the relevant contexts and in the various cultures they are easily understood by the receiver. A little cautionary note. You are more likely to come unstuck with these in different parts of the world where, if the emblem exists, it can mean something completely different from what you intended. You could end up with:

- a village bride
- a pack of mules
- a black eye
- or all three – if you're on a roll!

3 Affect displays

These are movements that tend to *give away* your emotions, positive or negative, and are usually subconscious. These include **facial expressions, gestures associated with the limbs, body posture** and **movement.** We'll be talking a lot about these as they reveal much about how we are feeling – to other people and also to ourselves. They constitute the *'leakage'* that in many cases we'd rather hide.

4 Adaptors

Similar to affect displays, adaptors are a *mood indicator* and are difficult to consciously control, making them a good barometer of someone's true feelings, be they positive or negative. They indicate whether the person is telling lies or engaging in a more serious form of deception. Adaptors include switches in posture and movements (alter-adaptors), actions that are directed *towards* the body such as rubbing or touching the face (self-adaptors) and actions like chewing a pencil, removing spectacles or fiddling with jewellery (object-adaptors).

 *Adaptors include switches in posture and movements* 〞

5 Regulators

These are movements related to our function of *speaking* or *listening* and also indicators of our intentions (we'll discuss 'intention movements' later). Head nods, eye contact and shifts of body position come under this category.

BODY WISE

As you work through the other Lessons please remember to interpret everything from *two* angles – with both 'hats' on. Don't forget – you're a receiver *and* a transmitter of body language.

Always ask yourself two questions:

1   What signals are people sending out (that I need to *decode*)?
2   What signals am I sending out (are they what I *intended*)?

Let's just accept from the start that:

● what people say is quite often at odds with what they *really* think or feel
● as a receiver of information, it's up to you to **interpret** the body signals to ascertain the *true* meaning of the message.

This can significantly affect the outcome.

When you communicate with others you need to know:

● if you are sending out good **positive** body signals
● that as a sender of information, you have the self-awareness to recognise and eliminate any **negative** body language that delivers the *wrong* message.

This can significantly affect the outcome.

Conscious or subconscious?

So, before we carry on, just a recap about the role of body language in our personal interactions with other people. As well as spontaneous behaviour, body language is something that we can use *purposefully* to influence an interaction.

All of us go about our daily lives transmitting messages to the world through our body language. But remember these two points as you go along:

- Some of these gestures are deliberate (and therefore **conscious**).
- Many gestures are beyond our control and are due to our physiology (and therefore **subconscious**).

Therefore, when we look at some body language movements we can see that they fall into the category of **voluntary** and **involuntary** movements that are governed by our thoughts – which produce emotions.

In some instances they give us valuable information relating to a person's thinking and, therefore, the feeling being displayed through a bodily movement. However, not being an exact science, as we shall see through the 7 Lessons, sometimes these non-verbal gestures and movements may just be *responses* to a stressful situation. That alone, in many instances, is also **valuable** information from a *communication* point of view.

What to look for – the Big Two

Let's simplify things right at the outset with some key points. These are the two things you want to be aware of at all times during any interaction – you want to know whether the people you are with are showing signs of:

- **comfort or discomfort (or anxiety)**
- **open body language or closed body language.**

Use this as a 'shorthand' for reading body language from now on – and for ever! Please commit it to memory.

This shorthand process will help you immensely because whenever you're with people, your newly trained eye will immediately focus on these two points, which validate each other:

- Do I detect comfort in this person's demeanour – or discomfort?
- Is their body language open – or is it closed?

Open or closed?

The clues as to whether a person is comfortable or not would be validated by **open body language**. Discomfort, meaning any kind of negative state such as *anxiety, fear, nervousness* or *hostility,* would be validated by **closed body language.**

So already we've taken a huge stride in recognising a person's emotional state. Of course, it's not the difficulty of being able to decode this state that's the problem for most people. It's usually laziness or lack of awareness – *or both.*

❝ *We've taken a huge stride in recognising a person's emotional state* ❞

TRY IT

From today make a point of sharpening up your mind-reading skills by *looking to really see* and *listening to really hear*. Begin by trying to recognise open and closed body language with everybody you come into contact with.

We've spoken about gestures and the need to base interpretation on clusters rather than trying to work things out from one solitary signal. In virtually every encounter you have with other people it's of immense importance to observe the clusters that signify either open or closed body language.

Of course the terminology speaks for itself when we consider our everyday language. Who's more welcoming? The person who says, 'I'm open to offers,' or the one who says, 'This is non-negotiable'? The boss who says, 'Pop in anytime if you have a problem, my door is always open,' or '. . . my door is always closed!'?

- Open body language is **welcoming, relaxed** and **attentive**. It signifies a lack of barriers of any sort, be they physical or extending from your own body. Your body is open and exposed and you're suggesting that you're vulnerable to others but you're comfortable about it. **Your hands are usually in view, possibly with exposed palms, which signifies submissiveness, and your legs and posture are free and easy, and eye contact is good**. Everything indicates a positive state of affairs.

- Closed body language is a cluster of gestures, movements and posture that **brings the body in on itself**. If you're experiencing the 'fight or flight' situation when you're threatened in some way, the tendency is to make the body appear smaller and to look for barriers to shield you from the threat.

Bringing the limbs **close in to the body** achieves the closed effect and a **barrier** can be put up by crossing the arms. This closed position is often used when you want to show that you're not a threat to the other person (some people of a more introverted nature may adopt this pose) as well as showing when you are uncomfortable in the situation you are in, or being with a particular person. **Little eye contact and tense shoulders and limbs that are crossed (folded arms and legs) typify this negative situation.**

Take a moment to consider what we've just discussed. What's your 'signature' position. **Do you exhibit these two types of body position in different situations?** Bet you do.

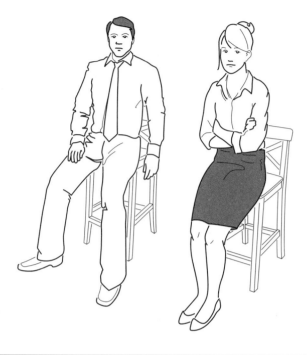

TRY IT

Put yourself in a closed body position. Note how it changes your mood. The mind affects the body – *but the body also affects the mind*. Now adopt an open position. See how your mood changes.

Observe other people in these two positions and note how you perceive them – and their message. Is it deliberate or subconscious? Can you think of people in your life (work or personal) who display these two types of position? Does it affect how you respond to them?

BODY WISE

'Open' body language is welcoming and relaxed, whereas 'closed' brings the limbs close in to the body.

Displacement activities and self-comfort gestures

These are the main sources in our quest to be able to read people better. We look for activities that reveal the clues to a person's state of mind – and accordingly how the 'relationship' may develop. But we cannot take single gestures in isolation: this is where people often come unstuck.

“ We cannot take single gestures in isolation ”

Gestures have been likened to a single word in a sentence. **You can't make meaning or sense from a single word, but when words are put together with others to form a sentence then we have meaning.** It's like that with body language. We piece together a number of clues that may point to the same thing. This is why we talk of clusters. Identifying these leads us to a certain conclusion.

So can we really say just because a person touches their nose when asked a certain question that they're lying? That when someone shifts position while sitting that they're feeling nervous? That folded arms indicate boredom with the listener? Or that interlocking ankles are hiding aggression? **Of course not.** As single, isolated gestures they indicate nothing. If all of these actions occur during an interaction with someone (a cluster of signals) then there's a good chance that there is a negative attitude in this person and so it might be a good time to change tack and/or try to get to the root of the dissatisfaction.

It could be you, your message or the environment (context) that is causing the problem. Many people mistakenly go through life thinking that they're good at reading people's non-verbal signals. They'll take a single action by another person and, lacking the requisite empathy to dig further, they'll ascribe it to a particular feeling – with no back-up information. Needless to say, these people may end up antagonising others (*'No I'm not bored with what you're telling me; I'm just tired' . . . 'No you're not – you're bored, I can tell' . . . 'Will you just leave it for now. . .'*).

So you need a lot of information to make a judgement about a person's attitudes. It's one step along a path on which you're looking for clues. Hasty and incomplete information leads to poor readings.

BODYtalk

Q I'd just like to ask a general question before we move on through the Lessons. You know when you meet somebody for the first time and they appear trustworthy and friendly. And then with someone else you get the opposite – you don't trust the person as soon as you begin a conversation. Is that because of body language?

It may not be consciously apparent to you at the time, but your 'gut feeling' or intuition is telling you that whatever is being said by both people may be perfectly okay, it's just that with the second person you're getting mixed messages. Their verbal and body language signals don't 'mesh'. Some kind of involuntary signal, from their face, their posture or a gesture, sends you a subconscious message that you don't feel comfortable about them. We'll explore all this later.

Q Are we saying that the brain is capable of seeing beyond the reach of the traditional five senses, into people's minds?

Let's put it this way. In any interaction with another person, your brain takes in a vast amount of information from both their body language and the vocal aspect of body language – 'paralanguage'. The senses send back this information to your subconscious where it manipulates 'data' (for want of a better term) received from your life experiences, to form that 'gut feeling' or intuition we just spoke about. It then sends this to your conscious brain which makes a decision as to how you feel and respond. It all happens in an instant.

Q Are some people naturally better at picking up signals and 'reading' people?

Yes. Just like some of us are naturally better at playing a sport or a musical instrument, or singing or dancing.

But it doesn't mean we can't learn these skills. We may not match up to the 'gifted' few, but we can certainly be proficient – we may just have to practise more. And of course you know what happens the more we practise something – suddenly we're good.

Q Does that mean we're all going to be 'body language wizards' at the end of these 7 Lessons?

Of course you can be – and I hope that's what the out-come is. If you start really seeing and really listening – backed up with everything that we'll have covered – you will sharpen up your own self-awareness, and then you'll certainly see the magic.

Q So we've got to remember some figures at all times, is that right? 54 . . . 30 something . . . ?

Don't worry – it's okay. It's 55, 38, 7. You will remember these, I guarantee, by the end of the final Lesson. Just be aware of the reason why many of us 'fall by the wayside' in our relationships with other people. By the way, when we talk about relationships we're not just talking about social and 'affairs of the heart'. We're talking about any relationship whether it's with acquaintances, working relationships, serv-ice providers or business clients – you name it. We all have the capacity to attract or repel people.

Q So this 'first impression' phenomenon that we're always being told about is something we should take seriously?

Well it's not a phenomenon – it's an instinctive dis-like and distrust between one person and another, and the information that decides this is communicated in a very short space of time. Your brain computes an awful lot of information in an instant.

Q I think we've got the message. Body language is all about picking up information on a subconscious level, and that tells the truth better than any words. Is that it?

Couldn't have put it better myself.

Coffee break . . .

 Since body language is a window to a person's mind, we need to have empathy to enable us to be aware of their feelings, as well as sincerity and perceptivity.

 We have to be aware of three essentials in order to read people accurately: context, clusters and congruence.

 Failure to interpret these correctly will always result in an ID 10T error.

 It is essential to control and be aware of your own emotions as well as assessing the emotions of others.

 If you're trying to get inside the mind of the other person by observing what's happening on the outside, remember they're doing the same to you.

 Be aware that your own gestures might be provoking a reciprocal gesture from the other person (not good if it's negative).

 The science of body language – not surprisingly when you're dealing with human beings – is not an exact science.

 Feelings are communicated more by non-verbal language than with words.

 It's no exaggeration to say that we go around in life either attracting people or repelling them through our display of body language.

Body language will always be the most trusted indicator for displaying our feelings, attitudes and emotions.

 If your words are not congruent with your body language, even though it might just be a bad habit (rather than an indication of your feelings), it's the fact that you might be misunderstood that's important.

 Research surveys consistently show that body language powerfully communicates the following:

- acceptance and rejection
- liking and disliking
- interest and boredom
- truth and deception.

 These are just some of the ways that we communicate:

- vocal
- facial expression
- spatial distance
- eye contact
- posture
- touch
- dress
- hand, arm and leg movements
- body tension.

 Over 90 per cent of meaning in any communication comes from visual body language and vocal elements, with the rest from the actual words.

So you have to be aware of:

- how you look
- how you sound
- what you say.

 If there's a 'mismatch' or lack of 'congruence' we'll believe the higher figure.

In reading body language you should always be looking for signs of comfort and discomfort and observing whether the body is open or closed. These are the foundation stones of your interaction.

Lesson

2

'When two people meet and make eye contact, they find themselves in an immediate state of conflict. They want to look at each other and at the same time they want to look away. The result is a complicated series of eye movements, back and forth.'

Desmond Morris

Looking

We're going to concentrate on the face and eyes in this lesson because it forms the centre of our non-verbal communication. The face is second only to the eyes, in body language terms, in revealing information about us. But our facial expressions are, for the most part, under our control. It's quite easy to change the face to reflect any emotion. If you want to look happy, even though you're feeling gloomy, you can put on another 'face'.

It's harder to control our gestures and vocal inflections – there's more chance of non-verbal 'leakage'.

People tend to believe what your face tells them rather than the words that they hear. The focal point of the face is the eyes. They reveal the *most* information, followed by your facial expressions. So, no matter what subterfuge you get up to in order to try to conceal your true feelings, a momentary grimace, narrowing of the eyes or a raised eyebrow can blow your cover.

We communicate more with our eyes than with any other part of our anatomy. In fact, I can see impatience in the eyes of a number of you as they glaze over – you're anxious to get on with the Lesson. You, of course, could sense how I'm feeling too, from my eyes. Communication is two-way traffic.

❝ Communication is two-way traffic ❞

In film and television you've probably been struck by how telling or powerful a particular scene proved to be by virtue of the 'expression' in the person's eyes. The great director Alfred Hitchcock said, when discussing the impact of dialogue and body language in his films:

> Dialogue should simply be a sound among other sounds, just something that comes out of the mouths of people whose eyes tell the story in visual terms.

BODY WISE

More communication is conveyed through the eyes than any other part of the body.

Eye contact

Eye contact is one of the non-verbal ways to:

- **express liking/intimacy** and show how the relationship is progressing (we look more at people we like than those we dislike)

- **exercise control** (for example, we may increase eye contact when we're trying to hammer home a point or be persuasive)

- **regulate interaction** (the eyes are used to direct the 'momentum' of a conversation, after having initiated it in the first place)

- **provide mood and character information** (such as attentiveness, competence, credibility, liking – as well as disengagement).

It makes you wonder how we ever manage to communicate with e-mail and telephone, doesn't it ?

The mirror to the soul

Leonardo da Vinci referred to eyes as 'the mirror to the soul'. Experts are still trying to work out what the enigmatic Mona Lisa is conveying with those eyes, but in our own everyday interactions we can perform the task much more easily. If you think about it, when we converse with people we spend most of the time looking at their faces, and so the eyes play a vital part in revealing inner thoughts and emotional state.

Eye contact plays a big part in striking rapport with people and establishing trust. Lack of it can seriously hinder your message – as well as preventing you from using this important body language signal to gauge a speaker's sincerity.

At a basic level we tend to look in the direction of what is of interest to us (be it a person or object) and look *away* when it's of little or no interest. From that premise we're provided with an initial clue to our own and other people's feelings.

There is no question that eye contact plays a significant role in being able to read other people because it is the behavioural trait that we notice immediately. When the appropriate eye contact is displayed we find a person to be – initially at least – trustworthy.

Eye behaviour

The term 'gaze behaviour' is used when studying the psychology behind eye movements and how appropriate the activity is in various situations. We know that in a normal conversation eye contact tends to be *intermittent*. So let's look at the behaviour we adopt in our culture when we're conversing with people. Any deviation from this may hinder trust and liking and may make us feel uneasy.

The eyes are so powerful that a gaze held for just a few seconds longer than the 'norm' can give out an extremely powerful signal.

- A speaker will look away from time to time and then go back to eye contact to make sure the person is still **listening** (and that they haven't fled the scene!) and to check from their eyes that they are still **interested** and **understand** the gist of what is being said.

- A listener displays that they are **interested** in a conversation by looking at the person much more frequently.

- If **confused** about something that has been said or if they **disagree** with it, or when **distracted** or just **bored** by the encounter, the listener will make minimal eye contact.

- If the listener is **looking away** all the time then something serious has occurred – a complete breakdown of attention.

In a normal conversation there is a 'dance' of gaze shifts:

- I start my conversation and glance **towards** you.

- When the momentum of my words takes hold, I **look away**.

- As I come to the end of that particular point, I **glance back again** to check the impact of my statement.

- You, as the listener, have been watching me and are now ready to **take over** as speaker, so you in turn begin your conversation, **look away** and then **return** to check the impact of your words . . .

So, in a typical interaction, this is a common pattern with the eyes going back and forth to check the listener's reactions to your words – you're watching their body language and they're watching yours (as well as making sense of the words).

For some of us, eye contact seems to be a difficult thing to come to terms with. Given that in the western world it is so important in establishing trust, lack of it can severely hinder your progress with others. Think about how you feel when you're dealing with somebody who averts their eyes when they're speaking to you. (Of course, we've all at some stage – or regularly – avoided eye contact with someone, as we pass them in the street or in a store perhaps, because we didn't want to talk to them.)

❝ *Eye contact seems to be a difficult thing to come to terms with* **❞**

CAUTION

When you fail to make eye contact, you come across as some-body not to be trusted as far as the message from the other person's subconscious is concerned (*'I don't know. . . something not quite right about him . . .'*). As always – something we'll be coming back to repeatedly during the 7 Lessons – it doesn't matter if, as far as you're concerned, this is unfounded. You may be perfectly trustworthy; it may just be your personal-ity type; or a hard-to-shift habit that's a 'hangover' from your childhood. The point is that it's the *impression* it creates.

BODY WISE

Regardless of a person's real reason, lowering the eyes and not making normal eye contact is not a confident gesture but is perceived as a *submissive* one.

Where to look

Many people feel uncomfortable with eye contact and are not as good as they could be because they – for whatever reason – find it difficult to know **which area of a person's face (or body) to** look at. Experiments have been conducted on three levels and it's culturally accepted that these are what can be classified as **'normal'** behaviour.

1 Broadly, in say a situation with strangers – including business – the ideal gaze is achieved by imagining *a triangle on the other person's forehead with the two corners of the base coinciding with each eye.* Don't allow the gaze to fall below the other person's eyes. What happens is that the act of lowering your gaze takes it closer to the social level (see below) *and so alters the formality or seriousness of the discussion.* Of course, if you reach a perceived level that enables you to (and because you want to) make it a more friendly encounter, then this is permissible.

2 On a social level, the gaze falls below the other person's eye level and is best described as the triangular area *between the eyes and the mouth.*

3   On an intimate level, when men and women want to show that they are interested, the gaze traverses the eyes and goes *below the chin* to other parts of the body, say down to the neck in the case of a man looking at a woman. **Usually at this level it is a discreet scan downwards before quickly returning to the eyes.** *But it's long enough to know that interest is there.* So you'll find that both men and women will use this gaze when **flirting** – and if the other person is interested they'll return the gaze.

So, train yourself to engage better with people and make eye contact at appropriate times so that you 'dance' with the rhythm of the conversation.

Oh, and just watch how much more you find yourself engaging with people and how *their* body language changes towards you. Keep intermittent eye contact for the right length of time, otherwise it descends into a stare.

Staring can make people uncomfortable and completely distort the 'message'. What is amazing is how many people do not realise that they have a habit of staring – they're probably never told about it and carry on in blissful ignorance. You may know these people socially or at work, or they may be acquaintances or, of course, strangers that you come into contact with. They make you feel uneasy.

So we assign them honorary membership of the – ever-growing – *'I don't know – there's something I don't quite like about him . . .'* club.

The 'dance' of gaze behaviour

So, of course, we're saying that eye contact is essential when speaking to people, but it will interest you to know – and I say that because it has probably, up until now, never occurred to you – that *the speaker looks away more than the person who is listening.*

Is that right? **You've probably never thought about it even though you've been interacting with other people for years.** You simply haven't noticed. (Just like you've never noticed which way the Queen's head faces on a £1 coin; or which way the pips face when you cut open an apple. Anybody know?)

So what's this all about? We're telling you to look at people when you speak to them and now we're asking you to disengage eye contact. Well, the reason is that – delicate things that we human beings are – we find too much gazing uncomfortable. If you use too little of it you give the wrong impression – nervous, untrustworthy. Use too much and you could appear aggressive or 'weird'. So you'll see from the figures below that the accomplished speaker, with high emotional intelligence – or empathy, to put it another way – will spend on average **around half the time or slightly less making intermittent eye contact.**

We look away to concentrate on our thought processes and be free from the visual distraction of the person with whom we're communicating. People who think that this is a sign of rejection or rudeness have not appreciated the frailties of the human condition somewhere along the line. We look away to 'give space' to the person to whom we are talking and it gives us **fewer sensory stimuli** to deal with. In addition, it's within the realms of a polite gaze without descending into a stare. Of course, if you spend your time alternating between looking at a person and scanning around the room, it can give the impression of boredom or disinterest.

BODY WISE

In any normal conversation eye contact is always *intermittent*. You feel uneasy about people when there is a deviation from this 'norm'.

Effective eye contact is also used to enable you to indicate to the listener when you're about to **finish**. If you observe people in conversation you'll see that subconsciously they'll make brief eye contact with the listener as a signal to indicate that it's their turn to speak.

What's also interesting is observing people who are waiting to speak, but perhaps can't get into the conversation because of a verbose speaker or someone who lacks empathy and can't read the subtle body language signals that the listener is giving out. The listener will look away – breaking eye contact. They may also make a gesture and inhale deeply, fill their lungs and start talking.

- Research shows that during a typical interaction a skilled *speaker* looks at the listener for 45–60 per cent of the conversation.

- The *listener's* gaze behaviour amounts to 70–80 per cent.

- Around 30 per cent of time is spent on what is termed 'mutual gazing'.

- The length of time that someone holds our gaze is also pertinent. The average length of time for a gaze is 2.95 seconds; the average 'mutual gaze' lasts 1.8 seconds.

It's probably clear to you now that when we experience that intuitive feeling or gut reaction about 'discomfort' when dealing with someone, *it is quite often related to the amount of eye contact or the length of time that they hold our gaze when speaking.* The socially acceptable gaze differs from a *stare*.

BODY WISE

In a normal conversation the speaker always looks away more than the listener.

As to their intentions, we have to look for other body language clues to support a reason why you think a person is holding your gaze slightly longer. As Desmond Morris said when describing a situation at a party:

The problem is trying to tell whether the person who 'super-gazes' you fancies you or actively dislikes you.

So, is the person 'a mad axeman' or something more benign? Is there a 'romance' in the offing or reinforced steel doors?

All that the eye directions reveal is whether you're the recipient of slightly more attention or slightly less than the cultural norm. Other non-verbal clues are needed to decipher the signal with more accuracy.

❝ Other non-verbal clues are needed to decipher the signal ❞

So how much eye contact do you tend to use every day:

- socially
- at work
- in encounters with strangers?

Start analysing it just to reveal things you never knew about your own body language.

What about your work colleagues, your boss, your relatives and friends?

- Are they better than you – or worse?
- Do you find that you make more eye contact with some people than with others?

BODY WISE

People often indicate their interest in another person (for whatever reason) by holding eye contact for a few seconds longer than usual.

What about at work? Is it different for:

- subordinates
- equals
- people who are 'higher up' than you?

TRY IT

Make more eye contact (within the acceptable levels we've described) with people in different areas of your life. See if there's a noticeable difference in your interactions. Try out a little excessive eye contact with a friend (tell them first) and ask them to tell you at what point they felt uncomfortable.

Dominance

We all come across people who make it difficult for us to look them in the eye because they may themselves use *excessive* eye contact. This disrupts the fine balance or 'dance' as we mentioned earlier. Alternatively, they may not use *enough*. You've seen this type – they may spend most of the time looking out of the window or at the carpet and then suddenly cast a glance your way (*that's assuming you haven't already left the room*). You may come across this type at a party or function, or it might just be a dominant individual because of your unequal status at work, for example.

There's an interesting finding relating to this *dominant–subordinate* encounter. Trials have been done which show that in most cases when two people of *unequal* status are in conversation, the 'dominant' person likes to *show their higher status* by doing the *reverse* of the normal looking – listening – talking dance that we've described. They spend more time looking at the other person *while they are talking* than they do *while they are listening*.

Our 'gaze behaviour' is largely determined by culture, and we know how long we may look at another person. **All the studies show that people with eye movements that can be seen as relaxed and comfortable, but at the same time attentive, come across as sincere, caring and trustworthy.**

Eye contact is great for developing trust but it can be done to excess. Use your instinct. What makes you feel uncomfortable? Cultures vary, but in the west too much gazing (staring) is impolite, can be seen as threatening and will alienate people.

" *Use your instinct* **"**

Of course we need to maintain eye contact with other people for the simple reason that *we have to watch other people's body language* in order to gauge our feelings about them. We also need to see their reactions. Worth repeating again – it's two-way traffic.

BODY WISE

If you feel that eye contact is going to be important for a difficult conversation – say in a restaurant or at work in a meeting perhaps – then try to orchestrate it so you are sitting opposite the person, rather than to the side or right next to them.

Just a word about cars. What's that got to do with eye contact? I'd just like to bring up a point about this since many people spend a lot of their time in cars.

In your private or professional life try to avoid conducting difficult or important conversations with the person sitting **next to you** or **behind you** in a car. Doesn't usually work. You need eye contact in a front-on position. (*How many scenes of slammed car doors have you witnessed on screen and in real life?*)

It's not overstating it to suggest that the degree and level of effective eye contact, more than anything else, helps in contributing to establishing that magical 'Holy Grail' of positive interaction – **rapport** – a state that exists *between* people. How can we define this elusive state?

**Rapport, like truth, beauty (and contact lenses!)
is in the eye of the beholder.**

Research on eye direction

We know that as 'the window to the soul' we can tell quite a lot about what a person may be thinking from their eyes. There's been a lot of research analysis to discover if the *direction* of a person's eye movements can tell us whether they are accessing and creating thoughts in terms of *sounds* or *pictures* or *feelings*.

What does all this extensive research show? Well, imagine you're in a conversation with somebody and while they're talking to you or listening to you, the following eye activity goes on. You've seen it all the time and you do it all the time (no doubt subconsciously). These are the findings:

- If a person's eyes move to the **right** and also **down** then they're trying to access **feelings**.
- If the eyes move to the **left** and **down** then they're talking to **themselves**.
- If the eyes move **upwards** and to the **left** then they're trying to **visualise** something that happened before.
- If the eyes move **upwards** and to the **right** then they're trying to **imagine** something.
- If the eyes just move across to the **left** then they're trying to **remember** sounds.
- If the eyes move across to the **right** then they're trying to **reconstruct** sounds.

Blinking

What do we know about blinking and what can it tell us? We blink at an average rate of **8–15 times per minute** (depending on the *situational* aspect). From a physiological viewpoint we're designed to blink so that the cornea is well lubricated. If you're in front of your computer staring at the screen or watching TV you'll have a variable blink rate because of concentration. So abnormal blink activity is usually associated with discomfort of some sort.

Keep an eye out (excuse me again!) for a person's change from **normal blinking rate to rapid movements** – perhaps 30–40 times a minute – because this indicates anxiety of some sort. People put under sudden pressure will exhibit rapid blinking, and equally a person who may be lying will display this along with other telltale gestures. When a person returns to their normal state, the blink rate slows down accordingly.

66 *People put under sudden pressure will exhibit rapid blinking* **99**

You'll notice that some people actually **flutter their eyelids** if they're struggling to explain a point, or are embarrassed about something, or as a response to something they've heard that they're not happy about. That's usually due to **discomfort** arising from:

- their own actions
- their inability to express a point at that moment
- something they've heard.

BODY WISE

Always check the moment when a person's eyelids flutter and use that as the cue to change the topic of conversation.

It's almost folklore now, but you may have heard or remember former US President Bill Clinton's performance before the Grand Jury and his blinking rate while under extreme pressure. It should be stressed again that on its own excessive blinking merely tells us about *discomfort* being experienced by the individual. Because the action of lying or merely being questioned about something are both stressful instances **they may cause a higher rate of blinking.**

In the main, the pointers relating to blinking display a range of behaviours linked to nervousness or anxiety about lying; or even a feeling of superiority over the other person.

You'll often see on television (political interviews, for example), or in real-life instances in and outside of work, that a person who feels themselves to be in a *superior* position to another will blink slightly *slower* than the normal rate you observe.

Think back, now that you're aware of this subtle gesture. Can you recall in your mind's eye somebody doing this? Try to picture it. **The slower than normal blink means the eyes are closed for longer than they would be.** *Blocking you out of the picture for longer than normal.* (The snooty shop assistant, doctor's receptionist, newly promoted boss – coming back to you now?)

You'll often observe this where the person in question has some kind of 'power' over you. This power may often be magnified in their own mind and be completely distorted or unwarranted, but it makes them feel good (and makes up for their lousy salary). The person is blocking you out of their sight for longer than normal with a slow blink. Is their head going backwards too? Mmm . . . that's handy – they can look down on you as well (that's your cluster identified!).

On the other side of the coin you need to be aware of the situation in which a person's normal blink rate **reduces**, so that they're blinking *less* than normal. You see this at conferences, in the classroom, during lectures and public meetings and in general conversation. It can be due to boredom, disagreement with what's being said or general hostility – **the glazed eyes of listeners who want to escape, exhibit infrequent blink activity.** The astute speaker picks up on the negative body language and changes tack or does something to get the audience interested again.

But, as with everything to do with body language, we're looking for supporting gestures – a cluster, remember – to evaluate; not just one activity in isolation. A person who is concentrating and

41

is really interested may be holding good eye contact, with less blinking, because they're enamoured with you and/or what you're saying. This is a *positive* sign so always check for other cues.

Cut-off eye activity

Some people display a tendency to close their eyes while speaking. It may be a general habit or happen *only when they're answering a question*. The former MP Anne Widdecombe used to spend a lot of time speaking with her eyes closed (she's improved a lot recently). The eyes will close for a blink but remain closed for a few seconds before opening again. The process is repeated throughout a conversation.

Do you know anybody who does this? Is it endearing or annoying? Margaret Thatcher was prone to this eye activity in the early stages of her reign as prime minister, but latterly the image advisers moved in and she almost totally eliminated it from her interviews. In many cases it's the body effectively 'shutting up shop' to block an unpleasant encounter or pressurised situation that the person is under.

" It's the body effectively 'shutting up shop' "

In other cases it's simply a person's way of blocking out other sensory activity in order to *concentrate* fully on their own thoughts. But the important point is that to the listener it can be **irritating** and send the wrong message – whether it's done for adverse reasons or purely as a mannerism.

Again, look for supporting cues to ascertain the emotional state of the person (clusters). If other gestures show irritation, you can be sure that this is another subconscious signal that gives you some idea of the person's negative state. If it's done in isolation, with no other clues, it's probably just a mannerism (but nonetheless irritating to *you*!).

Other eye activity

Eye dip

Just as some people will perform a variation on the theme of normal blinking, by closing their eyes for a few seconds repeatedly, we'll often show our desire for disengagement by *dipping the eyes* in a downward direction. We'll do it sometimes to break off from a conversation when the going gets tough. Other times we'll do it as a signal to indicate that by breaking any kind of eye contact, the other person can take momentary control and try to haul us back into the conversation. Sometimes, you ask a person a question that they're reluctant to answer and they may dip their head and stare down towards the floor. This is often referred to as an *eye dip*.

This brings to mind Hannibal Lecter in his prison cell, addressing a probing question – in his inimitable way – to a disturbed Clarice Starling as she averts her gaze and looks down towards the floor:

I'm sure you won't find the answer by looking at your second-rate shoes, Clarice.

The Silence of the Lambs

Hannibal Lecter (Anthony Hopkins) studies FBI agent Clarice Starling (Jodie Foster)

Charmer as always. Never mind – pass the Chianti!

Eye shuttle

Another common eye movement is where you see somebody rapidly moving their eyes from *side to side*. The head remains still – it's just the eyes that move (reminiscent of Benny Hill in a scene with scantily clad women!). This is commonly known as an *eye shuttle*. It's essentially a 'flight' response. You may have seen it with a builder perhaps, when you're questioning them as they're about to leave – cheque in hand – as to whether there's any kind of guarantee; or an acquaintance you've buttonholed after recognising them at a conference.

The darting eyes are scanning for an exit or some kind of help (as well as betraying nervousness or discomfort of some sort) – they just want to flee the scene. It doesn't give a very good impression to the person witnessing the activity. But it can be another telltale sign.

These 'cut-offs' are sent out to us subconsciously and, as with much other body language communication, are picked up by us in the same manner. But it leaves us in a situation in which our irritation can result in us displaying and sending negative body language *back* to the sender. Just to reiterate again, it's two-way traffic – I interpret your body signals (correctly or incorrectly) and respond to you with my own, based on my interpretation. **Is it any wonder that so much miscommunication occurs?**

The person with the 'irritating' habit or mannerism (let's call her A) sends out vibes to the other person (let's call him B) which he interprets in an unfavourable way. So he, in turn, because of his irritation with her, *reflects* back negative body language signals to A. She picks up these negative body signals and therefore becomes irritated with B. Breakdown! Because it's all about perception.

❝ *Breakdown! Because it's all about perception* **❞**

You can imagine a parallel conversation which A and B have with a friend later:

> *A* (to her friend): 'There was something about him – that sort of irritated look, I can't describe it. He wasn't listening either.'
>
> *B* (to his friend): 'There was something irritating about her. And she had this scowl kind of expression – couldn't wait to get away.'

Doesn't it just show the two-way nature of non-verbal communication? *Cause and effect.* And, as we've said before, most of it is done in silence.

If we're interpreting cut-offs as a negative signal that relates to us personally (you think, for example, they dislike you or think you could 'bore for Britain') and we're wrong – because we haven't taken account of *context* or looked for *clusters* – then there's been **miscommunication**. The signals that were picked up could simply have been as a result of the person being in a state of 'overload' and just wanting to reduce their 'sensory' input and concentration, even for brief moments. Some people need to block out visual noise in order to concentrate clearly.

In some cases perhaps it's just down to nervous mannerisms. However, in an ideal world such people need to be aware of what effect they're having on others – that's key. In other words, a knowledge of body language!

Widening and narrowing

Another eye activity that we all engage in from time to time is a **narrowing of the eyes**. What does it mean? It usually indicates some kind of disapproval of something or a sign to show dominance, and has been likened to the effect of *looking through a visor*. It accompanies a lowering of the eyebrows and so can often be mistaken for a sign of anger. These evaluations may or may not be correct.

Notice that some people adopt this expression when concentrating intently, reading something, a report perhaps. Prior to this, for example, their demeanour and open body language may have been quite amicable. You could be tricked into thinking that the contents of the document or report are causing unease, anger or disapproval in the reader. But at the end of reading it, the comment may be, 'Yes, that's good – well done.' What's that all about?

This person, with a particular personality type, obviously adopts a pensive or menacing look while **concentrating intently** on something. The face becomes set in an expression of studied concentration. Think Basil Rathbone (Sherlock Holmes) studying a document forensically, an exasperated Dr Watson at his side.

The disapproving person, by contrast, employs these subtle uncontrollable signals as they read items in the document that cause them **concern** or **annoyance**. The outward signals look the same in both cases. One is a consequence of personality type and the person's intenseness towards situations, and the other is a bodily reaction that betrays a certain facial expression around the eye area. So, as always, look at the 'whole' person before making snap judgements.

Have you noticed that when you want to show signs of incredulity or innocence, or want to show that you're paying attention and are really interested (especially women), you'll open your eyes wide, at the same time raising your eyebrows? We know that large eyes are appealing – look at your feelings for tiny, helpless babies with their large eyes. Men will often, in surveys, cite women's eyes as the thing that they first noticed about 'her' (contrary to popular opinion!). So raised eyebrows and eyelids help rapport.

Facial expressions

♫ With one look I can break your heart/
With one look I play every part/
I can make your sad heart sing/
With one look you'll know/
All you need to know/
No words can tell/
The stories my eyes tell/
Watch me when I frown/
You can't write that down/
You know I'm right/
It's there in black and white/
When I look your way/
You'll hear what I say ... ♫

(Sunset Boulevard)

As Norma Desmond sang in *Sunset Boulevard,* your look can speak volumes. In body language terms the face is the most expressive part of the anatomy; as in any interaction it is the first port of call – we naturally look at the face. Our words are supplemented by our facial expressions. We give out more signals with our face than with any other part of the body, **as you would expect with 22 muscles coming into play on each side of the face.**

❝ *We naturally look at the face* **❞**

A person's face will tell us that the person may, for example, be angry. Other body language will then tell us what those angry *feelings* are *making* the person do. Are they waving their arms or are they trying to repress these feelings and displaying tense movements such as rapid finger tapping?

BODY WISE

It's been noted in many research findings that women tend to be better than men in displaying effective body language, especially with the face.

We discussed eyes earlier and we know that they can be particularly effective at conveying messages – women especially can say a lot just with their eyes. Think Dame Judi Dench and her priceless laser-like stare in BBC's *As Time Goes By* (usually directed at long-suffering Geoffrey Palmer) and also frequently deployed to a chastened James Bond in her portrayal of *M* in the 007 films.

We know what eye movements can convey and this, combined with the facial muscles and nose, lips, mouth and jaw, reveals a lot. People tend to believe what the face tells them rather than any words that are uttered. If we say we're suffering from 'the blues', you'd expect our face to have a downcast expression. If we say we're angry about something or someone, you'd expect to see an accompanying expression. Equally, if we tell somebody that we dislike them intensely with a broad grin on our face – the face would be believed. In this situation there is a mismatch (or lack of 'congruence') and so we'll believe not the words but the 'visual'. (**Remember your 55, 38, 7.**)

Much research has been done on the face. It was Charles Darwin's research that highlighted the importance of studying facial expressions and the face's display of a variety of emotions. It is universally accepted that, across cultures, there are six easily identifiable emotions:

- Happiness
- Sadness
- Surprise
- Disgust
- Fear
- Anger.

Take a look around you when you're next in the company of many people. On a train, shopping mall, walking in the street. Be aware of the different facial expressions that people are exhibiting at any one time. Practice makes perfect. You'll subconsciously pick up and store a whole collection of facial mannerisms which you'll draw on when you're exercising that wonderful thing that we call 'intuition'.

❝ Be aware of the different facial expressions **❞**

If you had to describe in one word what a person's face reveals, what would you say? In public places see if their facial expressions are 'congruent' or if there is a mismatch with what the rest of the body is displaying.

Talking of congruence – just an aside – I recently saw a promotional video of a new airport terminal in which there were lots of people standing in a queue and smiling. What's going on there? I've never ever seen anybody at an airport with a smile on their face, have you!?

TRY IT

At the next opportunity, if you're sitting opposite somebody in any situation, take a look at the number of different
expressions and contortions of the face that the person goes
through in, say, 10 minutes or longer. Quite revealing.

Practise your mind- and body-reading skills and try to work out
people's **'inner voice'** activity (when their own thoughts relating
to their own dialogue are producing the changes in expression).
Also try to identify those related to **'environmental'** activity – on
a train, for example, when somebody boards and reduces another
person's elbow or leg room.

What about you? How many of the six universally identifable
emotions have you experienced in the last couple of weeks?
And, of course, while you were at work, home or out, you
managed to **'mask'** most of those feelings successfully, without
anybody in your presence knowing – is that right? In all fairness
– probably not. Much of your own 'leakage' is not even picked
up by you, especially your below-the-waist activity. Luckily for
you, only the empathic individuals you came into contact with
saw through the mask.

Smile and the whole world . . .

It's worth mentioning at this stage the name of an influential
US psychologist, Paul Ekman, who has conducted extensive
research over the past three decades into facial expressions. Just
as scientists were able to map the human genome eventually,
he, along with his colleague Wallace Friesen, has effectively
mapped all the expressions that we're able to produce and what
they mean. They've provided us with a better idea of which
facial muscles produce which expression. As a by-product of
this, it's also given us more insight into the activity of lying
(we'll be looking at this in more detail in Lesson 5).

Interesting research by Ekman and Friesen in 1982 on the six identifiable emotions revealed the following results from people who took part in the research. When asked to gauge:

- **happiness** – there was 100 per cent recognition
- **sadness** – about 80 per cent recognition
- **surprise** – this proved difficult to identify because of its transience
- **disgust** – about 80 per cent recognition
- **fear** – about 80 per cent recognition
- **anger** – about 80 per cent recognition.

However, it should be said that when these tests are conducted they're usually based on simulated posed expressions. When trying to evaluate genuine **spontaneous** expressions the figures come out at levels that are just above chance. In daily life we're more used to exhibiting two 'shorthand' facial expressions – the two relating to happiness and sadness:

- **smiling**
- an expression that relates to feeling **downcast** or **miserable**.

Both are easy to recognise as the research shows.

Take happiness – **it's the only positive emotion**. As the statistics above show, without exception we seem to be hard-wired to recognise when another person is in a positive frame of mind. When it comes to smiling, much research has been done in this area. It's of particular fascination to psychologists and neuroscientists because it is a frequent device for 'masking', along with the straight face.

❝ *Take happiness – it's the only positive emotion* ❞

Smiling is generally considered to be the easiest expression that we are able to display – almost a turn-on and turn-off display. And don't some people – like politicians in interviews or around election time if there's a handy, photogenic baby around – prove this to us?

The reason why so much research has been done on the subject is because when we're in contact with other people a smile, as you already know, impacts on **other people's attitudes towards us and encourages positive interaction**. Smile and the whole world smiles with you, as the saying goes.

TRY IT

See if the 'whole world smiles with you'. Make a point of adopting a smile when you meet people. Notice the effects. Do they smile back? Do people warm to you more? Is it leading to better relationships? *Is there a warrant out for your arrest?* (You're doing something wrong here!)

The effects of a smile

Think how strongly you're influenced by people who smile – those in your social life, your working life, when you interact with strangers. Do you go back to shops because you're influenced by a smiling face? Do you avoid places because subconsciously you pick up the lack of a cheery demeanour from the staff?

Maybe you go to restaurants because of the management and waiters with their positive, friendly attitude, in favour of another one that actually serves better food. You just feel more comfortable in the former restaurant.

Types of smile

There are two kinds of smile – the *real* (or felt) and the *false*. It's something that everybody displays – it's an innate facial activity that we engage in when we're happy.

We could add to that a subdivision of the false smile – the kind of smile that we put on when we're actually *unhappy* ('put on a happy face' as the song tells us). We all engage in this at some time or other and a person with good empathy skills would be able to pick this up. You're smiling through some kind of adversity or disappointment and to reveal your feelings would involve explanation, embarrassment or prolonged interaction with the person.

❝ You're smiling through some kind of adversity ❞

The vocal inflections ('Yes, I'm fine, thanks') should give some clues, as would the fact that with this kind of smile the corners of the mouth tend to go *sideways* rather than upwards. The eyes, in addition, would show no happiness in them and there is an absence of any wrinkling around them.

Let's just go back to the mid-1800s for an interesting 'time travel' experience. You may recall the term 'Duchenne smile'. Mean anything to you? OK – for those who are not familiar with the name, it refers to an interesting French neurophysiologist Guillaume Duchenne de Boulogne. Duchenne was interested in the muscles of the human face and also intrigued by the fact that we would often smile when feeling miserable *because we were able to do so.*

So we could go around pretending to be happy, which meant that we were walking around with fake smiles, and so when interacting with other people we were able to deceive them. (But, *quid pro quo*, they were doing the **same** to us.)

Duchenne's task was to *differentiate* the genuine smile from a fake one with his vast knowledge of the musculature of the human face. (Time for the squeamish to look away for a moment.) In his early career he used a volunteer who suffered from palsy for his research, but later he collected the heads of people who had been guillotined and used these to analyse the actions of facial muscles after electronic stimulation, varying the placement of the electrodes across the face.

Duchenne smile

What was Duchenne's groundbreaking discovery? Two sets of muscles control the smile:

- The *zygomatic* major muscles – these run down the side of the face and are linked to the corners of the mouth. When you contract these muscles they **pull the mouth back,** the corners of the mouth in turn are **pulled** up, possibly exposing the **teeth,** while at the same time giving the cheeks a pumped **up look.** They pull the corners of the lips **up** towards the cheekbone. These muscles that run from the corner of the mouth to the cheekbone *are under our conscious control.*

- The *orbicularis oculi* muscles are those around the **eyes.** Movement results in a **narrowing** of the eyes as the eyes are pulled back, with the telltale 'laughter lines' or **crow's feet** appearing, coupled with a slight **dipping of the eyebrows.** Duchenne's discovery was that these muscles *are not under our control and therefore reflect our true feelings.*

So a genuine smile or 'felt' smile involves the eyes. As Duchenne put it:

> **The first (*zygomatic*) obeys the will but the second (*orbicularis oculi*) is only put in play by the sweet emotions of the soul.**

(Didn't somebody mention something about eyes and 'mirror to the soul' earlier?)

The genuine smile (giving smiling eyes) is thus signified by the facial characteristics above – and there are two other significant pointers:

- a slow *onset* and *decline*
- bilateral *symmetry* in the face.

With the fake (social or masking) smile – giving unsmiling eyes – you tend to get:

- an abrupt *onset* and *decline*
- bilateral *asymmetry* in the face.

Ekman's follow-on work: the 'real' smile and the 'fake' smile

Following on from the work of his early predecessor Paul Ekman discovered another aspect of a real smile – with a real smile the movements of the lips (from the *zygomatic*) are shorter than with the social (fake) smile. In common parlance (as far as body language vernacular goes) we therefore talk about genuine (enjoyment) smiles and fake (or masking) smiles. We now know that when we experience true pleasure or enjoyment, the brain and its physiology *combine* to produce a smile involving the mouth, cheeks and – crucially – the eyes. Conversely, a smile that just involves the *zygomatic* – which is under our conscious control – is what is known as a fake smile.

I don't like to use the term 'fake' smile, which many psychologists and other researchers in the field tend to do, because it technically includes *social* smiles. As we've established, in much of life this smile is used as a 'lubricant' and denotes the opposite of hostility.

❝ This smile is used as a 'lubricant' ❞

So it's useful to be aware that within this smile category (which has been given the term 'fake') there is the social smile and what we can term the 'masking smile'. *This (masking) smile is used to cover emotion rather than to display it.*

For example, we often try to hide a negative emotion of nervousness in a situation with a 'masking smile'. Quite often you can see what really lies underneath the mask if you attune yourself to pick up the 'microexpression' (see below) that occurs when the face relinquishes the masking smile and it fades away. This can be very revealing.

That said, of course it is useful in certain instances to know when this type of smile is masking an underlying concern or antipathy. Ekman continued Duchenne's work for us and developed, with his colleague Friesen, the Facial Action Coding System (FACS) to measure and identify various combinations of facial movements. As a tribute to Duchenne's earlier work and Ekman's subsequent research, a real smile is thus often referred to as a 'Duchenne smile'.

Microexpressions

The FACS research also introduced us to *microexpressions*. These are 'leakages' that occur in the face for a short duration. Because of the way that humans are 'wired', our faces tend to betray our true feelings because our emotions affect our physiology; and the other way round (as we'll discuss later).

Our face will always leak out what we truthfully feel. Since what we feel – for example, anger or fear – activates certain facial movements, *behind* a smile there may have been a fleeting microexpression which lasts *a matter of seconds*, but nonetheless can give away our true feelings, despite the smile.

Our face – as you know only too well – is highly efficient in displaying our inner feelings. In a split second it registers a message after activating the appropriate facial muscles. An *opposing* message from the brain to hide an expression comes too late and so the true expression leaks out for a second or so, before being cancelled by the counter-expression. Since we try to suppress these lightning expressions in a split second – which typically only sophisticated equipment and playback and freeze-framing would reveal – it takes a really astute and observant person to detect them.

If you are a regular viewer of BBC's *The Apprentice* series, you'll know that they have a 'post mortem' programme (*The Apprentice: You're fired!*) that features a panel and the ousted candidate. They often show freeze-frames of the candidates and their fleeting microexpressions that reveal their true feelings beneath the 'mask'. Typically they'd be asked in the boardroom, for example, if the choice of project manager was a 'good one'. You'll see that after the nodding and the smiling, a fleeting microexpression appears on the face of one or more of the team, displaying the true emotion felt! Difficult to spot at the time (even for Lord Sugar) but very revealing in the playbacks and freeze-frames.

Put on a 'happy face'

Of course we've instinctively known through all our life's experiences that some smiles are 'felt' and some are 'masking' true feelings. We see it in people all the time. How many times have you 'put on a happy face' for an occasion when you've felt quite the opposite? (In fact, how many times just today?!)

I'm sure you've noticed at award ceremonies like the Oscars or BAFTA – those occasions of triumph and tears – the wonderful 'acting' that is displayed by disappointed nominees as the results are announced. Painful – and I'm not talking about for the nominees, *but for the viewers!*

You know what I'm referring to – the broad effusive smiles as the disappointed souls loudly applaud the person who has deprived them of an ornament for the mantelpiece or downstairs loo. Okay, Angelina Jolie, Gwyneth Paltrow – we understand your point. *What's a girl to do?* (Or George Clooney for that matter – guys put on the same 'performance'.) **It's just that . . . now, I might be being unfair, but you get the feeling that if they'd put on a performance like that in the actual film, well . . . might they have actually won . . . an Oscar or a BAFTA?**

Mixed emotions at an award ceremony

You put on this smile as part of your 'role', in your social life, with strangers, at work – just about with everybody you meet.

We don't go around trying to evaluate whether this sort of smile is genuine or not. In most cases it doesn't matter. It's a social 'lubricant'.

“ *You put on this smile as part of your 'role'* ”

Incongruent smile

We know that a smile has to be 'congruent' (one of the big 3 Cs) with the words or vocal content of our message. If we see a mismatch then our subconscious immediately focuses on that all-important word – trust. **We don't believe the message.** In the world of politics, because we live in a media-driven society, we're constantly exposed to examples of this, and for the astute observer of non-verbal language it provides fertile ground.

Politicians are always being told, by their highly paid image consultants, to smile. A recent, fascinating example of this was the creation of Gordon Brown's smile. Unfortunately it was likened by political columnists to that of the Child Catcher in Ian Fleming's *Chitty Chitty Bang Bang* (a character created originally for the film by Roald Dahl).

The problem for Gordon Brown seemed to be one of 'smile selection'. It was used on a hit-and-miss basis, and was completely random. Loss of child benefit CDs – smile. Fuel prices at a record high – smile. Food prices in the shops soaring – smile. Messages that were not consistent with body language!

As we know, in body language terms it means that we're witnessing 'incongruence' – the facial expression does not *match* the content (words). There's no evidence of empathy so the meaning of the whole message is void.

Make it natural

Is there a general message? Sounds obvious – smile only when it feels natural to do so, and make sure that it matches content and doesn't conflict with events. Have you ever had a friend, acquaintance, boss, work colleague, customer or anybody of whom you've asked for something deliver negative news relating to their *refusal*, with an over exaggerated smile (a non-genuine one, which of course you now recognise!)? What message does it convey?

Firstly, there's a *mismatch*. So you don't believe the smile is a true reflection of their feelings or regret at them being unable to grant what you wanted. (*We don't smile when delivering 'bad news'*.) So your feeling is that the person had no intention of agreeing with you and hasn't even considered it in the first instance – and may also possibly be disguising some antipathy.

Secondly, this 'deceit' is *irritating*. A neutral expression with perhaps the occasional smile as a person refers to certain points during a conversation is *normal* behaviour – it denotes at least a hint of sincerity. In other words, that this is a considered judgement (rightly or wrongly on their part) as to why they can't grant what you had asked for – or at least that they understand your point of view.

So if you're guilty of doing this, understand how, in body language terms, you're sending out incongruent messages that completely dilute the point you may be trying to convey. Plus you're possibly irritating and confusing the listener, and losing their trust. If your gestures are not 'congruent' with what you are saying and also with the rest of your body, there's a mismatch – and your 'image' suffers.

❝ Incongruent messages that completely dilute the point ❞

The social smile

The smile is probably the easiest of all facial expressions to identify. It's denoting a positive emotion and we display this when we're feeling happy. But it's also an expression that we have to adopt for social purposes. We're usually pleased that somebody is smiling at us – whether it's genuine or not – because most of us use smiling to indicate (as our ancestors did) friendliness as opposed to hostility.

Equally, people in jobs in which they are dealing with people – service industries, for example – are urged to smile as part of the job, even if they don't feel inclined to do so. So it's an expression that we've been used to displaying from an early age. The fact that we have to turn it on at will means that because of all the practice we've had, we're quite good at producing smiles that are not quite genuine.

This social smile is identifiable by:

- the two corners of the mouth moving *sideways* towards the ears (there is no lift upwards) and *no activity or emotion around the area of the eyes.*

The true (or felt) smile is identifiable by:

- the two corners of the mouth being forced *upwards* towards the eyes with the outer edges of the eyes displaying that familiar crinkling.

I can see through your smile...

Over the centuries the smile of the Mona Lisa has captivated poets and onlookers. It's accepted that the expression on a person's face is usually recognised from the two distinct areas of:

- corners of the eyes
- corners of the mouth.

With this painting, Leonardo da Vinci chose to let these two distinct areas fall into shadow and so there is confusion as to the mood of the subject of the painting and constant discussion relating to what the smile is telling us.

The point about identifying the two types of smile is that although we naturally engage in felt and social smiles (the latter – as discussed earlier – because of social convention and also to display friendliness), being able to *distinguish* between the two is extremely helpful. For example, let's say that at a party your host has invited someone you know they do not particularly like (out of duty). Compare their smile when they greet them to when they greet somebody you know they really care for.

That gives you a pointer on their genuine smile versus the fake smile. So now you know how to distinguish feelings in your own dealings with them. Transpose that to a work situation, for example. You know when your boss, colleague or whoever is genuinely pleased about a suggestion or is just being polite (she'll find a way to ditch your idea later). Change tack if you get this warning signal.

" *Change tack if you get this warning signal* **"**

It's self-evident that you would want to know whether or not someone is genuinely experiencing these positive emotions towards you so that you can respond appropriately. **In addition, knowing and recognising the two types of smile in a conversation means that you can pursue the elements of a conversation that evoke genuine interest and investigate (or ditch!) those elements that display the opposite smile.**

It's a pity that – despite exhortations from employers to their staff – many people don't smile enough! Repeated studies show that we are attracted to people who display genuine smiles (which we perceive subconsciously) and regard them as sociable. We are affected by people who smile at us even if we don't realise it at the time. We'll discuss this shortly when we talk about the groundbreaking studies relating to smiling and emotions. But before that here's a good example of its effects in a 'service' situation that we're all familiar with.

Many studies have been done in the USA (the home of tipping!), in test conditions, involving waiting staff and the 'gratuity' received by pleasant and smiling staff compared with those who didn't smile.

- Without exception, the ones that smiled and displayed friendly body language collected more tips.

- The waiting staff who knelt down to floor level to discuss the menu with seated diners seemed to do even better. (Posture and movement are powerful body language actions.)

- It was also shown that waiting staff who used the power of touch – say gently touching the customer on the elbow or side of the arm – seemed to strike up a greater *rapport* with them and this again influenced the giving of tips.

There are many types of smile and, just as we've spoken about open and closed body language, you won't be too surprised to learn that there are open smiles and closed smiles. Kate Middleton has a wide open smile showing emotion around her eyes (charming all around her!)

- Open smiles tend to show the teeth

- Closed smiles don't – think of a tight-lipped smile, or an asymmetrical or 'lop-sided smile'.

Kate Middleton

Smiling and emotions (chicken or egg?)

We mentioned the name of Paul Ekman and his pioneering work earlier. He's provided us with invaluable information regarding findings that were discussed decades earlier. Modern equipment has enabled scientists to prove the link between facial expressions and emotions and the effect on our **autonomic nervous system.** Ekman and his staff at the University of California have studied facial muscles in depth.

His work ought to – in a perfect world (but the world *not* being perfect and all that) – ensure that those of us who go around with:

- a scowl *or*
- a disconsolate 'visage' or other 'negative' facial expression (sometimes by choice and other times through simply not being *aware* of how our facial features are arranged)

should stop immediately. Why? What's the deal?

Well, it was always thought – and taken for granted – that as the face is the barometer for our emotions, then facial expressions follow *after* an emotion has been felt. So it was assumed that you first of all feel happy, or sad, and *then* the corresponding facial expression is displayed.

Now wait for the great news – and be sure to tell those morose shop assistants, surly waiters, bank clerks, theatre ticket clerks, doctors' receptionists, security guards, supermarket check-out staff, railway ticket staff, office receptionists, your work colleagues, your boss (if I've missed any of the guilty out there, humble apologies – *you know who you are!*) – that:

- If you force yourself to display a certain expression, the mind and body work together and physiologically recognise and process that emotion, with appropriate *biochemical* changes.
- So if you are feeling tense and unhappy, for example, and are told to change your downcast facial expression to a smile, *this makes you feel better inside because of the feel-good hormones that are now working their way through the body.*

Magic! So the research has turned things upside down. Can the expression itself create the emotion? Yes.

❝Can the expression itself create the emotion?❞

Ekman's six emotions (detailed earlier) were tested for ANS (autonomic nervous system) changes, which control heart rate, breathing, body temperature and other functions.

● Startling bodily changes were observed with *heart rate increase* and *raised skin temperature* for the negative emotions, like sadness, disgust and fear. The ANS activity was highest for anger.

● In the lab experiments these bodily changes were not present when the facial muscles were changed to a *smiling* expression. In fact they served to calm down the negative ANS activity when the subjects were asked to switch to a smile.

Political columnists were forever highlighting Alastair Campbell – former Downing Street adviser to Tony Blair – and his perpetual scowl and look of anger as he went about his business. With a face almost permanently set for conflict and anger we can see how the ANS would have made it self-fulfilling, reinforcing the emotions.

TRY IT

Even if your facial expression doesn't match how you *actually* feel (for example, a permanent angry look even though you don't feel that way inside all the time), lab tests on your ANS would show an increased heart rate, raised skin temperature and other body hormone stressors. (Lighten up!)

Lips

We've seen how the position of the lips in relation to the smile is governed by the main facial muscles. These muscles are versatile and **work with or without each other**. That's why you can have an asymmetrical or 'lop-sided' smile. One side tells one story about your feelings, the other tells another – **pleasure and pain**. A good example of a lop-sided smile, tuned to perfection, is shown by actor Harrison Ford when he's not cracking the whip. The lips are also very expressive in giving away our feelings quite *independent* of a smile. They're a good indicator of all our emotions.

Much has been said about open body language – the more open the mouth the more relaxed you appear. Conversely, tense or tight lips on a person show some kind of restraint of an emotion – usually a negative one.

Pursed lips are common with many people. Usually this signifies somebody thinking intently about something but they're not ready to speak, either because it's not possible for them to interject yet, or they haven't carefully thought out their argument. It's almost certainly a sign of *dissent* so if you spot this during a conversation, try to intercept and get to the bottom of the disagreement.

Victoria Beckham's signature expression

Sometimes we'll bite our lip (as in the saying) – we'll cover this in Lesson 5.

The pout seems to have come back into fashion thanks to former Spice Girl, Victoria Beckham. Lips pressed tightly together with tongue pushed up to the palate. Unlike the pursed lips, the pout seems to cover a multitude of sins as far as feelings go – sadness, anger, disgust and even *imaginary* ones. If you see it – look for cluster behaviour. Otherwise it may just be for effect. After all, it's popular in Hollywood.

❝ *The pout seems to have come back into fashion* **❞**

The position of the head

We'll be looking at the powerful use of head 'nods' as a communication device in listening behaviour in Lesson 3. What about the position or angle of the head in various instances? Take a look at these and consider the signals that they send out to other people (whether they're right or not).

- **Completely lowered head** (with slumped shoulders) – sends out a message about your state of mind through where you're looking. You're *looking* down, so therefore you're *feeling* down. It can denote that you're feeling depressed, guilty about something or that you're concealing something or that you're fatigued or extremely tired.

  I've noticed a new phenomenon over the last few years – it's been evident in many settings. For example, as I walked into a restaurant once, there was a group of people – nine or so – around a table and some of them were talking animatedly, while others were in a slumped position with their heads down and their hands under the table. It looked odd – as though they were on the wrong table. Well, I thought, maybe it's deference – *they're shortly expecting a visit from the Pope.* No. How about shy, maybe? Lacking social skills? Depressed? Then, as I was shown to my table, I could see that they were all tapping away at their smartphones. Their posture had given a completely wrong signal to an onlooker. (Quite apart from that, isn't it rude anyway?).

- **Slightly lowered head** – you may be trying to avoid eye contact (for whatever reason). It's sometimes adopted when you're showing respect to somebody. Sometimes we adopt this as a form of protection from something – possibly because our space is being 'invaded' in a crowded environment.

- **Tilted head** – often used to show interest in something that's been said or about to be discussed. Other people display this to denote their curiosity about something. At the other end of the scale it sometimes may be a 'shyness' gesture; equally it could indicate that a person is doubting something that has been said or implied and is looking for reassurance from the speaker. If you're in this situation and observing these body language signals then try to look for any 'cluster' behaviour that gives you a clue to what signal is being sent out. For example, if a head tilt is accompanied by the sudden folding of arms and a perplexed expression – then something's wrong.

Many people are unaware of their style of walking posture or sitting posture and it's only when it's pointed out to them that they take notice of their habitual style. The same is true for the position of their head.

❝ It's only when it's pointed out to them that they take notice of their habitual style ❞

A simple maxim – **look straight ahead and you look to the future. Look down and you're probably feeling down.**

Always remember that there's a biofeedback mechanism that occurs with our mind and body (see more in Lesson 6) – remember the effects of smiling on the body.

BODYtalk

Q Even though the eyes are so expressive, we do need to give out other information with the face to back up how we feel, don't we?

Of course. When you look at somebody, the eyes can convey love, liking or hostility. Pupils contract when a person is angry or in a negative state so that this, in conjunction with the muscle movement around the eyes, makes the eyes look less friendly. But it's the facial expression that accompanies it that denotes the feeling – a smile for liking or attraction, or tight lips and jaw for dislike, for example.

Q Sometimes when I'm with people they're looking over my shoulder or around the room between any eye contact they manage to make. I find it irritating. Should I be annoyed?

Certainly you should. Unless they're on the run from the FBI or anxiously awaiting the arrival of someone (which they should inform you about) there's no excuse. Whatever words they may be saying are all invalidated, I bet.

Q I work with someone – I'm actually in charge of her. She's very nice and very competent, but she always seems to lower her head and dip her eyes when we're talking. A bit like Princess Diana used to do. What do I make of this? Am I doing something wrong?

Well, she's giving you the opportunity to take a dominant role – any dip of the head is what we'd call a submissive gesture. I understand you're in charge of her, so you have that authority anyway. It often signals a desire not to be involved in the interaction in the first place. It's usually used by women and it does, no doubt, inspire you to feel protective towards her. It's used a lot when women are flirting because it makes the eyes seem bigger and evokes innocence.

People respond to large eyes in a protective way (think of babies). It may just be her 'style' of communicating when she's with you. If you're not being manipulated – and you can handle it – it's fine.

Q I need your help on this. I've got a kind of serious face, I suppose. I think people may get the wrong idea when I'm talking to them. I try to strike up empathy but my face doesn't show it. Any ideas?

People believe what they see rather than what they hear. If there's a mismatch they'll believe and take meaning from the higher figure (remember 55, 38, 7). So you're obviously, with the best will in the world, trying to convey with your words (7) how you feel, but there's no 'congruence' – one of our 3 Cs. So they believe what they see – your facial expression; and sorry to be blunt but it obviously turns them off.

Q What can I do?

Well, you know about the importance of a little smile. You don't have to do a full-wattage Julia Roberts. Just occasionally, at the right moments, back up your supposed empathy with a change of expression. I'm sure it will work wonders.

Q Reading what you just said above, our faces are obviously honest in showing how we feel, but we can, to a certain extent, control our expressions. My question is, how does a person see through the masking?

As you know, we're always looking for 'congruence' and 'clusters' – two of our 3 Cs. See if the expression agrees with the vocal (the paralanguage) – any speech abnormalities or tension. Also with the visual clues – any cluster of body leakage that negates the single piece of information that we have, namely the facial expression. If these cues are informative, then the masking is revealed. It's quite easy to control our facial expressions (if you don't want to look sad you can fake it). It's much harder to control our tone of voice or our gestures.

Coffee break . . .

 The face is second only to the eyes in revealing information about us.

 Since we communicate with the eyes more than any other part of the anatomy, it follows that eye contact plays a big part in striking up rapport with someone.

 Eye gaze for strangers (and business) is normally directed at the triangular area between the forehead and the base of the eyes.

 Eye gaze on a social level is normally directed to the triangle between the eyes and the mouth.

 In normal eye contact the speaker always looks away more than the listener.

 Normal eye contact is always intermittent (if this doesn't happen it makes us feel uneasy).

 Lowering eyes and avoiding eye contact is not a confident gesture and is perceived as a submissive one.

 A person often shows interest in another person by holding eye contact a little longer than usual (a few seconds more).

 We need to maintain eye contact for the simple reason that we need to be observing other people's body language, quite apart from everything else.

We know that large eyes are attractive (just ask men and new mothers), so to show you're interested and paying attention, try to adopt a wide-eyed look (with raised eyebrows) rather than a narrowing of the eyes.

 The face may reveal a person's mood, but it's the other body movements which reveal what those feelings are making the person do (for example, waving arms or tapping fingers restlessly).

 All the research shows that women tend to be better than men at reading body language generally, especially from the face.

 The six easily identifiable emotions are:

- happiness
- surprise
- disgust
- fear
- sadness
- anger.

 Research suggests that we seem to be hard-wired to recognise happiness, the only positive emotion, more than the others (100 per cent in tests).

 Two types of smile are generally referred to: the real (or felt) smile and the fake smile. We could add a third if we wanted to make a distinction, the unhappy smile, when we're putting on a 'brave face'.

 The true smile engages the *zygomatic* muscles, which run down the sides of the face, and the *orbicularis oculi*, which are the ones around the eyes.

The *orbicularis oculi* muscles are not under our control so the brain and its physiology provide an indicator of a genuine smile.

 The other pointers to a genuine smile:

- Bilateral symmetry in the face (as opposed to bilateral asymmetry for the fake).
- Slow onset and decline (as opposed to rapid onset and decline for the fake).
- Smiling eyes (as opposed to unsmiling eyes for the fake).

The lips are a good indicator of our emotions regardless of a smile. Generally the more open the mouth, the more relaxed and attentive you appear because tight lips suggest the harbouring of a negative emotion.

Lesson

3

'There was speech in their dumbness, language in their very gesture.'

William Shakespeare

Listening

We spend most of our lives listening. Your relationships – and the quality of them – are, for the most part, determined by your skills in listening. In fact, in the whole communication process it's your skills as a listener that determine your effectiveness.

During any interaction there's a dual status that you occupy. At times you're either:

- listener *or*
- speaker.

If you're typical of most people (hand-on-heart time) you prefer to be talking rather than listening. As Larry King, the former US television talk show host for CNN, once put it: *'everybody's fighting for airtime'*. Noted for his listening 'style' as he interviews celebrities, politicians and businesspeople every day, he also commented that many other interviewers prefer lecturing rather than listening.

In our own everyday lives the majority of us love to hear ourselves talk and don't really care to listen unless it involves 'us'. Of course there will always, fortunately, be exceptions and you probably find yourself quite attracted to these people.

❝ The majority of us love to hear ourselves talk ❞

Active listening

When a person is speaking to you – and you occupy the role of listener – **do you show with your body that you are listening?** That you're present? That you understand what they're saying (even if you may disagree)?

People who engage in what has come to be known as 'active listening' not only listen, but also are *seen* to be listening.

The elements of any speaker's message comprises of:

- the words spoken
- the body language (visual)
- the non-verbal 'paralanguage' (auditory).

As we stressed earlier, of course words are important, but people make a decision about you and your message first and then decide whether or not to stay around and continue with any interaction. This applies to life and all its relationships. The message is usually interpreted through the visual body language and *listening between the lines*.

Visual and vocal cues

What do we mean by listening between the lines? It's tuning in to the **vocal** aspect of body language – in other words, **pitch, tone, volume, rhythm, rate of speaking and all those paralinguistic clues that reveal more than the words themselves** (we'll be looking at this later in this Lesson).

Time and time again, all the surveys show that the most charismatic, successful – or just plain popular – people are great listeners; and, more importantly, they *show* it. How do they do it? **Through body language.** Their empathy shines through and they're sensitive enough to know when to speak and when to listen – and, more importantly, they *show* that they're listening with their whole body. Result – rapport.

These people are also empathetic and look to see *beyond the words* that are spoken and listen 'between the lines'.

They tune in to that second element of non-verbal language that makes up – along with the visual – more than 90 per cent of the meaning in any message (the 38 per cent, remember?). **How you say the words – that vocal element.**

We typically listen to the words that are being uttered but fail to tune in to the emotional meaning. If you listen with all your senses you'll be more attuned to engaging your 'sixth sense', or 'intuition' or 'gut feeling', call it what you will. We'll be talking about this *paralanguage* later in this Lesson.

> ## TRY IT
>
> The next time you're listening to somebody face to face, try to suspend your own thoughts and don't think about formulating your reply. See if you hear and remember more.
>
> Then, when you've graduated from that, try to train yourself to listen to the 'paralanguage' – the *way* that things are said. Eventually, it will become your listening 'style'.

It's all too easy to blame the listener or 'audience' in a meeting or social setting when they disagree, or – in your eyes – have missed the point. Your 'performance' has not struck the right chord – **you didn't pick up the body language signals that suggested doubt, uncertainty or hostility on their part, and so your lack of awareness precluded you from even trying to rectify the situation.**

Listening with all your senses

Listening and responding in a way that helps you to understand another person's perspective – and at the same time shows that you are truly listening to them – is the first stage of establishing *rapport*. Yet it would not be unkind to suggest that most of us are poor – no, let's be generous here – terrible listeners.

❝ Most of us are terrible listeners ❞

Equally, it would not be an exaggeration to say that for the majority of people, their lives are ruined by poor listening – and that goes for listening to words said, as well as the way that they're said and the all-important body language that accompanies it.

We all have a tendency to 'switch off' or 'drift off' if we're subjected to a bout of listening without being able to talk ourselves. Yet it's important that we listen using all our senses since we know that the true message is often not relayed through the **words** that are being spoken.

There is a developmental condition on the autism spectrum known as **Asperger's syndrome** and those afflicted have great trouble listening and also picking up on the body language of others. Like all autistics they have poor social skills and so there is a difficulty in interacting with others. Their inability to read facial expressions and their poor eye contact means they are oblivious to the feelings of others. The situation is made worse by the fact that the lack of ability to pick up on the vocal element of body language, tone of voice, means that the different nuances and meaning of words – based on the tone of voice used – is completely lost.

Hearing and listening

Most of us – and it's a problem that stems from childhood confuse *hearing* with *listening*. Time to make you feel a little guilty now. Are you aware of the difference? Well, let's start by saying this:

- one is a *physiological* process
- the other is a *psychological* process.

Hearing is an auditory activity in which the sensory process through the ears makes a journey to the brain – a physiological approach. Listening involves the interpretation and understanding of a message after it's been through the hearing process – a psychological activity that makes sense of what's been heard. The two processes work together to give meaning.

It also means that it's possible to hear something without actually listening to it. You know the situation well – it probably happened to you at school. You're busy daydreaming with half your attention given to the teacher who's talking about 'stalactites' that grow from the ceilings of caves, interrupting your thoughts about last night's episode of *Sex and the City*. The teacher, noticing your slouched posture, asks you to repeat what she just said and you reply, with a startle: 'stalactites grow from the ceiling'.

Phew . . . Out of trouble for now, but you didn't take in the meaning, you just recited it while it stayed in your short-term memory and it will dissolve in the next 50 seconds or so – to be forgotten. There's no psychological activity of true listening and making meaning and processing, and therefore 'storing' the information. You were merely 'hearing'.

So, I think we've established that listening is something we all take for granted. In reality it's not that easy setting aside the concentration even to listen to the words. But, as we know, it's important to listen not only to the words, **but also to how they're said** – listening to all those paralinguistic cues. Also, we have to observe what we see – **and show that we're listening**.

'Listening' body language

So let's have a look at 'listening' body language:

- **making good eye contact**
- **using head movements**
- **mirroring (in a natural way) body language**.

We've spoken about eye contact in the previous Lesson. You're well aware of the unwritten rules of eye contact etiquette. Eye contact helps a speaker to be confident that they're being heard and that you're taking an interest. Nobody likes engaging in a conversation with a person whose eyes are continually darting around (cocktail party style). That's why people who are good with their eye contact are perceived as more likeable and interesting.

Head movements are an interesting area in relation to encouraging people to speak and generating rapport. It's mainly through the 'head nod'. Five different 'yes' types of nod have been identified:

- the encouraging nod ('Yes, how fascinating')
- the acknowledgement nod ('Yes, I'm still listening')
- the understanding nod ('Yes, I see what you mean')
- the factual nod ('Yes, that is correct')
- the agreement nod ('Yes, I will').

The bowing action of the nod appears to be an inborn action, just like the head shake. The head shake, signifying no, is thought to stem from our time as babies when the negative response to being spoon fed is to turn the head first to one side and then to the other.

❝ *The nod appears to be an inborn action* **❞**

Many people fail to develop rapport with others by not showing through the body (usually a head nod) that they are fully engaged. As we saw above, there are five possible messages we can convey to the speaker. It's a simple action and as a gesture performed by the listener it can help the conversation to flow.

Lack of nodding (not listening with your body) can stifle a conversation because the speaker may think one of two things:

- You're not paying attention to what they're saying.
- You're not interested.

BODY WISE

Studies show that listeners who engage in repeated nodding activity tend to elicit as much as *four times* more information from a speaker compared to when there is no head activity.

A tip: Check the direction of gaze that accompanies the nod. If they're looking *away from you*, it usually means that they are ready to start speaking. If they're looking *at you*, they're just conveying their agreement with what's been said.

If you've watched experienced chat show hosts on television, you'll notice that there's a lot of nodding activity as an encouragement for the guests to 'open up'.

What's the message from the 'head nod'?

Generally, **the speed of the head nod** indicates what the listener is conveying. Many people are confused about this – and have never bothered to delve into the meanings. Let's get this straight now, because it will really help in your future interactions and avoid confusion – and also *annoyance* on the speaker's part when you don't get the message!

- **Slow head nod** – is usually an *encouragement* nod to get the speaker to carry on talking. (They're also indicating to you – the speaker – that they don't want to switch roles yet.)
- **Slightly faster** – they're telling you that they *understand*.
- **Very rapid** – either that they totally *agree* (arousing emotions in them) or possibly that they want to interrupt and become the speaker.

You, as the speaker, need to check for other bodily clues as to which of these is the one that the listener is trying to convey.

Jonathan Ross encourages Sir Bruce Forsyth to open up

If you don't currently use these head movements it would be worth trying them out in your future interactions. See what difference it makes. You'll find that conversations will last longer, be more open and that the 'turn taking' will be much more natural.

While we're on the subject of head movements during listening and how people show that they're interested in what you're saying, you've come across what's known as the '**head tilt**' during conversations (we discussed them briefly in Lesson 2). If we go back to Darwin, for a moment, his interpretation among humans and animals was that it was a 'non-threatening' head movement, which indicated interest in something.

If we roll the years forward, studies show that, of course, he was right – we tend to do this subconsciously when we're listening attentively because something has caught our interest. You'll see it in audiences in cinemas, theatres, in meeting rooms at work, during training sessions and, of course, in conversations with all and sundry. **Like the head nod, it is a submissive gesture.** It's thought that for some of us it recreates the feeling we had as a baby when resting our head against our parent's body, when looking for comfort or rest. Think for a moment of someone you know who uses this head tilt. What feelings does it evoke in you? Do you use it yourself sometimes? Consciously or subconsciously? I'm sure the answers are very revealing.

❝ *It is also a submissive gesture* **❞**

Mirroring or synchronising (to build rapport)

You've probably come across the term 'mirroring' when it's used in the context of interpersonal activity. It does cause confusion. Think of it, for now, as being in synchronicity with another person – when you're getting on well. **Being 'on the same wavelength'**. You don't mirror the other person's body language exactly. You make yours similar to theirs and you try to adopt their general posture, in a completely natural way.

After a while, your body language instinctively becomes similar to the person you're with, as do the 'vocal' aspects like rate of speaking and loudness of voice. You don't try to match another's paralinguistic style exactly. The head nodding becomes 'in sync' and other postural movements and hand gestures all seem natural and follow a similar rhythm. You lean forward to show you're listening. Eventually they do the same. There's a rapport through body language.

'My kimono is open'

We instinctively tend to do the opposite – in other words, display negative body language – when we're experiencing discomfort or don't agree with what somebody is saying, or when perhaps in a work/business interaction we're listening to somebody who speaks in jargon or corporate buzzwords with irritating clichés – 'My kimono is open', for example, meaning I've disclosed everything (I'll just give you a little time to recover from that!) We've been through much of the 'pushing the envelope' and 'blue-sky thinking' phase, but new monsters keep regenerating from MBA schools and the like. But if these words stop people listening then it hasn't enhanced communication – it has impeded it.

Either way it stops people listening, and so it's the reading of 'listening' body language that alerts the speaker that it's time to change tack, and also maybe explain what something means, before completely losing the audience. We may sit further back in our seats, for example – as opposed to leaning forward. This can cause a similar reaction in the other person. So you have matching or 'mirroring' – **but of negative body language**.

If you see this happening, be aware of your own emotions and **get back to an 'open' posture.** If you try to resist closing up and continue to adopt an open style then there's more chance of a favourable outcome and less chance of them staying in a 'closed' body language position. **Non-verbal language is quite contagious – for better or worse.** If they're stuck in that mode, hand them something to look at to unlock their arms.

Vocal aspects of body language

'It ain't what you say' – it's the paralanguage. The vocal part of body language relates to the 38 per cent that we spoke about earlier. We can vary tone in a number of different ways:

- The *pitch* of the voice can be varied – we can go from high to low and move between these two levels during a conversation.

- The *speed* of speaking can be varied – we can speak rapidly or at a slower pace.

- The *loudness* of the voice can be varied – going from soft to extremely loud.

- The *rhythm* of the voice can also be varied.

The term for this non-verbal aspect of speech is **paralanguage**. The *para* comes from the Greek denoting 'alongside' or 'above and beyond'.

Let's talk about *pitch* first because it's capable of expressing a wide range of different meanings. So its importance as a part of non-verbal body language is huge. For example, it's easy to see how the pitch can indicate a contrast between making a statement and asking a question – 'They're coming back already' as opposed to 'They're coming back already?' We could add a third – '*They're* coming back already?' We can signify a bored state by using a monotone, or we can express surprise by increasing the pitch.

❝ We can express surprise by increasing the pitch ❞

The *speed* or tempo, as it's sometimes called, can indicate different meanings. We know that something expressed at rapid speed indicates urgency of some kind, while slower or more deliberate speech conveys an altogether different meaning. It can be indicative of a person's state – nervousness or insecurity, for example. There are of course differences in personality types resulting in some people adopting a more impatient style of speaking, which is quite rapid in its delivery. Equally, at the opposite end of the spectrum there are perhaps the more 'introverted' types who may be more guarded in their delivery.

Loudness is another aspect that conveys different meaning. Generally, we regard a very loud voice as conveying anger. This brings to mind the booming voice of an actor I once saw in a play – playing an actor – and his constant refrain to his wife, in that booming voice was – 'I warn you. I'm in a mood.'

These three aspects of speech together provide a *rhythm* to the voice. **People with attractive voices have a package that proves pleasing to others** (more about that later).

So it's plain to see that unless we've had the benefit of being trained in voice production, as actors inevitably have been, we don't tend to give that much thought to how we sound – and how others are perceiving us. If you're losing out in the working world on job interviews, promotions, effectiveness in meetings or selling a product – or just about anything – then could it be that there's something you're not being told? Maybe you don't sound right.

Along with breathing, all the elements that we've discussed contribute to clarity and good diction. If you never pause for *breath*, for example, it can irritate and grate on the nerves of the listener.

- If you *breathe deeply* from the abdomen, it gives you a more *relaxed and confident* sound.

- If you *shallow breathe*, with shorter breaths, it's because you're *nervous or anxious*.

BODY WISE

A simple maxim: *how you breathe is how you sound.*

Remember that good posture is essential for good speech. **Hunching your shoulders, or just generally slumping, is not good for your delivery; neither is tension in your throat or stomach.** Before that telephone call, or before you walk into an interview room, breathe in slowly and deeply through the nose – make sure you hold for a few seconds as you inhale – then release the breath gently through your mouth.

Voice is obviously important to us in our liking and disliking of people. Think about radio presenters, specifically those DJs that provide chat and play songs as part of their radio show. They're unable to provide us with any visual body language. We determine our liking, in the first instance, from the voice. Was BBC's long-running Terry Wogan show so popular simply because of the songs he played? Then there's the legendary Tony Blackburn recently back on BBC radio (and reunited with award-winning radio presenter Phil Swern). Few decades on and still pleasing listeners. Or John Humphry's on Radio 4's *Today*? Is it just the music or the news? No, we subliminally know which voices are pleasing to us.

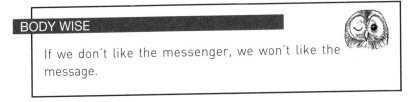

BODY WISE

If we don't like the messenger, we won't like the message.

So, if you think you're doing everything else right and you can't understand where you might be going wrong, check the non-verbal element of your voice.

Formula for a perfect voice

Research from the linguistics department of Sheffield University in May 2008 – in a paper entitled *Formula for a perfect voice* – produced some interesting findings. The researchers were trying to create a new formula for the perfect female and male voices. The formula represented the subtle blend of **tone, speed, delivery, words per minute** and **intonation**. Analysing the highest scoring voices, sound engineers and academics hatched a mathematical equation for elements that the ideal voice should contain.

Researchers found that the ideal voice should utter no more than 164 words per minute (wpm) and pause for 0.48 seconds between sentences that fall in intonation. The result was that a *combination* of Dame Judi Dench, Mariella Frostrup and Honor Blackman make up the perfect female voice. Dame Judi speaks at 160 wpm breaking off for 0.5 secs between sentences; Mariella Frostrup speaks for an average of 180 wpm and pauses for 0.5 secs; Honor Blackman articulates herself at a more considered 120 wpm.

How did the men fare in the findings? The most appealing voices were a *combination* of Alan Rickman, Jeremy Irons and Michael Gambon. Jeremy Irons talks at 200 wpm; Alan Rickman at 180 wpm; and Michael Gambon at 160 wpm.

Researchers also concluded that the vocal traits associated with positive characteristics such as trust and confidence scored higher – and so these produced the perfect voice.

As was said earlier – it's not just what you say ... *it's the paralanguage.*

BODYtalk

Q I'm not a bad listener, I don't think, and after what you've just been telling us I think the problem must be that I don't *show* that I'm listening. People don't tend to share information with me and I'm not having much success at work either. Surely, if I sit there and maintain eye contact, that shows I'm listening?

Not really – a stuffed dummy with well-manufactured eyeballs (and clear irises) could probably match you by the sound of it. It's not enough just to 'receive'. The 'transmitter' wants to know that you understand what they say, that you're actually 'still in the room' (not 'running your own tapes' in your head) and whether they have agreement or interest. Is that too much to ask? Wouldn't you expect the same in the opposite situation?

Q Yes, I suppose. How do I look enthusiastic then?

As we said earlier, you have to show you're listening with your whole body. A good listener is worth their weight in gold. People perceive those who listen to them in a very positive light. Try nodding because it encourages people to carry on talking, and the slight head tilt (common with women) shows you're paying attention. Lean forward in your chair to show open body language – this often encourages the other person to do the same. Obviously facial expressions, showing empathy at the right moments, confirm that you are paying attention. Right? Now see how your interactions with people change.

Q The matching of body language during an interaction . . . I think you also called it mirroring. How does that work? I know what you said, but isn't it a bit false?

Only if you misunderstand the purpose of it and forget its naturalness. Studies have analysed people in a state of rapport (you can use your own terminology for this if you like, call it 'getting on well' or 'on the same wavelength', it doesn't really matter). What matters is that when movements and speech eventually – and naturally – get to a stage when two people are almost mirroring vocal and non-verbal aspects, there's a good state of rapport. When you have that, the relationship flows – there's a synchrony or rhythm with the other person's actions and gestures.

For example, you'll pick up on the speed at which a person normally speaks, the way that they gesture and their seating posture. When the subconscious picks up that nothing 'jars' then movements and conversation flow. So if your tendency is to speak loud and fast, for example, and the other person speaks more slowly and softer, you adopt that style in a natural way.

Q I was really interested in that section on 'paralanguage'. I'm not that great with words and articulation. If I sharpen up on my – what I think you psychologists call – 'impression management' skills in presenting myself and spend more time on that, I should be okay. That's given me hope. Am I interpreting that right?

Err . . . (excuse the speech disfluency) no. ID 10T error. You have to understand one thing. If you don't get the visual body language right a person is not going to 'stick around' even to hear your words. If you get the visual right – okay, well done for that – but then when you open your mouth you're an absolute turn-off, you're finished. So it's not a substitute that we're after. We're looking for congruence – words and visual matching up to create the right impression.

Q I agree with what you've been saying. I know that sometimes when we interview women for PA jobs they're perfectly dressed, their greetings body language is good and they seem controlled when they first sit down – and then when they open their mouth it's downhill all the way.

Exactly. I'll just read out the conversation that the head of personnel for a city bank – who was looking for someone for a front desk position – had with her boss:

'One girl we spoke to had one of the most dreadful voices I've ever come across. She didn't vary the tone the whole conversation and so came across as someone lacking in energy or being very bored. When working on front line reception, as she would have been, that's not the impression we want clients to receive on walking in.'

Coffee break . . .

 Your relationships in life are determined by your skills in listening.

 Active listening involves 'listening with the whole body'; you need to show you're listening.

 Time and time again, studies show that the most charismatic and successful – and popular – people are good listeners – and are seen to be listening – and this conveys empathy and promotes rapport.

 Many people confuse listening and hearing because it's possible to hear something without actually listening to it. The first is a physiological process and the second is a psychological process.

 Good 'listening body language' is:
- making good eye contact
- head movements
- mirroring – in a natural way.

 Many people fail to achieve rapport because of the lack of any head nods and it can convey (sometimes mistakenly) that you're either not interested in what's being said or not paying attention.

Listening beyond the words spoken constitutes the 38 per cent of meaning (remember 55, 38, 7) that we derive from any communication.

The vocal aspect of body language is known as 'paralanguage' and refers to the pitch, speed, loudness and rhythm of the voice.

 If you breathe deeply from the abdomen you achieve a more relaxed and confident sound. If you shallow breathe – with short breaths – it's (whether you know it or not) because you're nervous or anxious.

 Recent studies have shown that vocal traits associated with positive characteristics such as trust and confidence score highly.

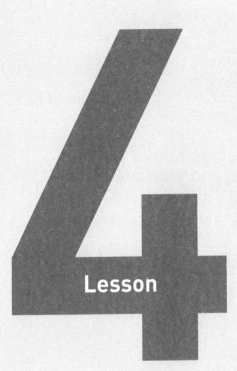

Lesson 4

'He that has eyes to see or ears to hear may convince himself that no mortal can keep a secret. If his lips are silent, he chatters with his fingertips, betrayal oozes out of every pore.'

Sigmund Freud

Limbs

We're going to look at the signals given out by our limbs in this Lesson. As you'll see, so much of our body language is conveyed through the **hands, arms, legs** and **feet**. We'll also look at the various *displacement activities* and *self-comfort gestures* that are associated with our limbs and how they leak information that gives away our true feelings. Please remember these two terms, which are widely used in body language vernacular. They form the basis of recognising signals that you may see and ones that you are sending out to the other person. They represent the key to your mind-reading skills – an indication of what internal thoughts are being processed by the other person (and you).

Hands

We'll begin with the hands, which are probably the most spontaneous of all our body parts. We use them to **greet, illustrate points as we're speaking** and generally **express the emotions that we're feeling**. We're hard-wired to use our hands in tandem with our speech. In fact the number of nerve connections between the brain and the hands is greater than between any other part of the body – the hands have 25 per cent and the arms 15 per cent of the total. The legs and feet we have less conscious control over – the further away from the brain, the less we're able to fake.

‹‹ *We're hard-wired to use our hands in tandem with our speech* **››**

A person's hand activity will communicate a lot to us and hand movements of our own will *influence* how we're being perceived by other people. **It's generally accepted that we can relay a message with greater clarity with accompanying hand movements.**

Status, hierarchy and dominance

If you watch people interacting with each other you can often tell the 'pecking order' as to how a person perceives themselves in terms of 'status' or 'hierarchy' or if they consider themselves to be on an equal footing.' Of course, it is observed more often in a working environment on an everyday level where you'll notice how touch sends out a message both to the person on the receiving end and to observers. In business and in politics, tests confirm that, in the main, the person who perceives themselves to be of higher status is the one using the hands to initiate the touch. You'll see it a lot in politics where the politician wants to let the audience (voters) know who's in charge – and make sure they remember it – as well as the person who's on the receiving end of the touch.

A wonderful example of this was demonstrated on the steps of 10 Downing Street in May 2010 when the two new coalition leaders met for a photocall to the world's press after the general election.

Prime Minister David Cameron guides his new Deputy into Number 10

Cameron and Clegg: what is their body language really saying?

(James Borg, *Daily Telegraph,* 15 May 2010)

. . .The symmetries are all part of their supposed political chemistry. On first glance, David Cameron and Nick Clegg are alike in so many ways, in appearance, age, height and education – and in their gestures as well. But exactly how alike are they in reality? Throughout last week, their respective body language gave subtle, unspoken clues as to the real state of their relationship.

Prior to the press conference, the pair posed briefly on the doorstep of Downing Street – a ritual that Cameron would have almost certainly have rehearsed in his mind. He certainly had finessed that relaxed, authoritative air. In these situations, and faced with a barrage of cameras, the winner is the person who can seem natural in what is essentially an unnatural environment. After the introductory handshake, we saw both men pat each other on the back, a signal which neuropsychologists refer to as a 'parental' gesture. And what do most parents signify with this movement to their young? 'I'm in charge.' It's a status reminder, and can be especially important in the 'first among equals' situation in which Cameron and Clegg now find themselves.

But their body language was more complex than that. Cameron patted Clegg first, who reciprocated with a pat of his own. Cameron then patted back, and Clegg did the same... before Cameron gave the assertive final pat with his right hand as he ushered his deputy through the door. This was both a classic repeat display of courtship, and a barely concealed power struggle. Crucially, by doling out the final pat, Cameron had the last word in the vernacular.

The press conference that followed was a chance to meet the 'newlyweds'. Cameron came across as more assured, more prime ministerial in his manner and delivery, making frequent references to his new partner by gesticulating towards him with his right hand. When he gives a speech, Cameron has a subconscious habit of splaying his fingers, an open-hand gesture that projects trustworthiness. This in stark contrast to the closed, clunking fist deployed by the previous resident of Number 10.

▶

Clegg, in between looking at his notes, attempted his now-signature delivery technique of looking straight ahead. However, with his general facial expressions more subdued than usual, he glanced down more than Cameron – a sure sign of nerves. After all, he had something to be nervous, and indeed embarrassed, about after being exposed earlier in the week as having been in talks on the sly with Labour – the romantic equivalent of an 'ex-girlfriend' – before finally deciding to go to the altar with the Conservatives.

When Clegg spoke, it was interesting to note that Cameron orientated his entire body towards him. When we are completely at ease and interested in another person, we turn not just our head but our whole body – and often the feet – towards them.

When Cameron spoke of the challenges facing his administration, Clegg turned only his head in his direction. He also displayed a number of microexpressions, fleeting subconscious gestures that last between three and five seconds, but which display discomfort. Clegg bit his lip on a number of occasions and touched the inside of his mouth with his tongue. This was noticeable especially when the subject of proportional representation was raised.

There was a change in Clegg's later demeanour. As the prime minister spoke, Clegg orientated his whole body and feet towards him – a noticeable shift. As the prime minister answered questions, Clegg began to give nods and respectful glances. Rather than implying complete agreement, this usually suggests something more crucial to a working relationship – deference. Clegg is acknowledging that, although he is now a powerful player, Cameron is very much the man in charge.'

Some people use their hands in a natural way while they are speaking; others find it difficult and so it takes a certain amount of focus and awareness to get into the habit. It's been noted that the more *extrovert* type of person will tend to use hand movements to accompany speech, as opposed to the more introvert type. You'll often see that the most effective speakers are those who display a lot of hand gestures to reinforce a point – and their timing is in synchronicity with their speech.

So, apart from the face, *our hands are the most expressive and communicative part of the body,* and as well as adding to our speech we quite naturally use them in place of words. Small wonder then that we use them constantly to reinforce a point – and also why they are responsible for displaying a lot of 'leakage'. So much so that we subconsciously pick up signals when we're not aware of seeing somebody's hands in a conversation, or even when we're observing them speak (if their hands are behind a desk, rostrum or under a table, for example).

Experiments are continually being conducted in this area of hand display, and time and time again the results show that *the perceived impression of somebody who is 'hiding' their hands is a negative one.* Now, this is not to say that this is always the reality – that hands are deliberately being concealed from view. *But impressions are formed by perception.* In fact, everything we're discussing in this world of body language is, as has been mentioned before, to do with perception.

How do you feel if your arms are hanging down by your side and you're asked to give a talk? If you're a person who uses your hands, however little, you'll almost certainly feel uncomfortable because your mannequin-like pose will make for an **unnatural** delivery and the visual appearance would not inspire confidence. So, in body language terms (remember 55, 38, 7) the impact of your 7 per cent (words) would be completely overshadowed by the way that you said it (38 per cent), and also what we see – your uncomfortable demeanour (55 per cent).

Palms of the hands: up or down?

In the field of study relating to hand gestures we'll look first at the position of our palms during our gestures. Much research has been done on the subject of palm position – to simplify, we basically have **up** or **down**. What's the significance?

Almost exclusively the *palms up* position wins 'hands down' (excuse me!). The findings show that when you use the palms up position when you're speaking, most of your listeners interpret the 'message' **positively**. Repeat this with *palms down* and the figure is much lower.

What's the reasoning behind this? We associate the sign of an 'open' hand with friendliness, honesty and trust. We take oaths with the palm of the hand – 'Can I be completely open with you?' says one of the parties round a negotiating table, as he gesticulates with his open palm thrust forward. Since it's usually a *subconscious* display, we tend to believe the gesture and it tends to put us in a more receptive mood to the other person's message. It's generally thought that people who are trying to fake a particular emotion or tell an untruth find it difficult to do so with their palms up.

Empty palms in ancient days equalled no weapon. That's how handshakes developed (more on that later). The point is, it's essentially a *submissive* gesture (think beggar on the street, or the villain's response in a black and white Sherlock Holmes movie as Holmes, Watson and the Inspector break down the door in the last scene). So it's a good indicator for you to judge the sincerity of a person when they're delivering their message. They're trying to convey to you that they are to be trusted ('please listen to me with confidence').

Now that's all well and good, I hear you say, but what about the estate agents, used-car dealers, liars and other purveyors of this gesture? Don't they know about this gesture and deliberately use it to hoodwink us? Of course they do. But by the end of Lesson 7 you'll be able to filter out this gesture if you notice that it's not 'congruent' (remember) with other 'open' gestures, and you experience lack of eye contact, an inappropriate posture, a false smile and a vocal give-away.

A gesture with the *palms down* is used to convey dominance or authority. You're issuing a command here. An example of this was Hillary Clinton early on in her campaign for nomination in the presidential elections in 2008, as she tried to impress on the assembled throng that she was in control. Later in the campaign her advisers must have asked her to change to palms up. She switched to a palm (singular) up position. (Note: people find it easier to break the previous palms down *habit* if they switch to a single-hand palm up rather than try to do a palms up with both hands.)

❝ You're issuing a command here ❞

So if our hands are so important when we're dealing with others, even if it's only subliminal on their part, then it means that in order to inspire trust and give out the right impression they need to be visible. However, we have to be careful because our hands are able to send cues to an observant listener who, as well as picking up positive vibes, may pick up negativity or anxiety.

You can just imagine, for example at work, a colleague of equal status to you gesticulating with a closed palm position, in an up and down fashion, and saying: 'Can you just move that pile of directories that somebody's just dumped there – they're in the way?' With that gesture, it sounds and looks like a command. It's not your boss that's asking you, it's someone of equal status. Result – could lead to antagonism. It's not the words, it's the gesture *and* the way that it's said. Double whammy!

Now suppose that same request (even with the same words) is made with an open palm upwards gesture. It then becomes more of a submissive request (or favour) or a sign of helplessness and is less likely to upset the other person. You are far more likely to carry out the request.

There's a variant of the palm down gesture which has all the fingers closed together and it's instantly recognisable as the adopted salute during the tenure of the Third Reich, seen many times on the grainy newsreels showing Adolf Hitler.

What's the point?

I'll just mention, while we're talking about the effectiveness of palms, an irritating gesture that occurs **when you point your index figure** at someone. You have the closed palm position, which turns it into a fist, with a finger point. **How do you react to a finger point?** Doesn't make you feel good, I'm sure. Evokes memories of childhood perhaps? Characteristic of a parent, teacher or other irritating person that we've come across in later life? It's an aggressive gesture and it's surprising how many people are unaware either that they do it or of the effect it has on the receiver. Almost throughout the world, it's associated with the same sentiment (*I think you get the point*).

If you are the proud owner of this gesture, attempt to get rid of it because it causes offence. It's okay for television perhaps – 'Natasha – you're fired' (*The Apprentice*) – but in real life it causes antagonism.

Displacement activities

Time for an important term in the psychological world of body language – *displacement activities*. These are things you do to help displace anxiety (or nervous energy, to put it another way). Please remember this term. It will help to serve as a memory-jog in helping you to *label* it and to *read* it in other people. These little actions can be fleeting, but very revealing.

❝ *These little actions can be fleeting, but very revealing* **❞**

Displacement activities are movements performed by us when we're **experiencing any kind of inner conflict, torment** or frustration. The essential point is that these are just *small* movements that may tell us an awful lot about what may be going on inside a person's head. As with most things in life, it's usually the *small* things that make a *big* difference. What small gestures do you make that 'leak' out your true feelings (usually without you even knowing)?

BODY WISE

If you underestimate the capacity of small things to make a big difference, then a quick question: *Have you never been to bed with a mosquito?*

Trains and planes

When an observation survey was conducted at railway stations and airports, where tension is often high, some interesting activities were noticed. Firstly, there were activities that were very clear in denoting tension. Secondly, there was a type of disguised activity because passengers obviously did not want to reveal, in any way, that they might actually be apprehensive – or frightened – of boarding a plane and that they would rather flee the terminal building.

Typically the 'displacement' behaviour involved constant rechecking of tickets and passports, making sure the wallet was still in the inside pocket (the reassuring tap), picking up and putting down luggage, gazing at a mobile phone repeatedly. All of these things had been checked, invariably only minutes before.

Interestingly, and to be expected, there was **10 times as much evidence of displacement activities at airports as at railway stations** (just 8 per cent of passengers boarding a train revealed evidence of these activities, while the figure was 80 per cent at the check-in desk at airports). In addition, it was observed that there was a lot of **head scratching, facial contortions, tugging of earlobes and general fidgeting and fumbling in the airports.**

There was also evidence of stress-smokers sitting quietly and making a big show of the whole smoking 'ritual' – locating the packet, finding the lighter or matches, lighting and putting out the flame, strategic placing of the ashtray, flicking of non-existent ash from the clothes and then finally the stubbing out of a barely smoked cigarette with an exaggerated action. (The extensive smoking ban means you'll now be unable to enjoy your role as a 'voyeur' and witness this elaborate display of inner conflict. However, you'll see a similar performance in *other* situations, no doubt.)

I'm sure this is going to make you all the more aware of your own behaviour the next time you're waiting to board a plane. Also it will raise awareness of your *own* display of displacement activity in many situations, and once you've identified them you can aim to rid yourself of these telltale activities if you're in a situation in which you don't want to come across in a particular way.

We all have our own personal displacement habits that come into play to rescue us when we're experiencing a period of conflict or tension – chewing gum, nail biting, eating, drinking, keeping a cigarette in the mouth (even if unlit). Movie star Clint Eastwood had an interesting displacement gesture in the so-called Spaghetti Western films he made. He would have the unlit cigarette in the corner of his mouth at all times, but instead of removing it from his mouth whenever he had to speak to someone (in what was usually a tense situation), he would *light* it and then start talking, with it still in his mouth.

The hands, because of their close association with the brain, are particularly prone to self-directed activities when we're experiencing conflicting emotions. These displacement gestures provide comfort and dissipate energy at the same time. They tend to fall into two areas – the first is **externally** directed; the second is **internally** directed (self-comfort gestures) – this is covered later.

Displacement that is *externally* directed includes fiddling with clothing or objects such as keys, pens, spectacles, coins, the stem of a wine glass, a mobile phone or rings on a finger. The point is that it may be a reflection of our feelings, but it's irritating for the onlooker. Look at the 'context' and see if you can identify 'clusters'. Then take action.

❝ It's irritating for the onlooker ❞

Smoking styles

Smoking (as discussed earlier) is another external displacement activity worth elaborating on because it gives us some additional clues. **When feeling anxious or nervous it gives the hands something to do.** Much research has been done into styles of smoking and what these display about a person's state of mind at the time – positive or negative.

What do we know? Well, barring situational factors – like proximity to other people and being in a confined space – that determine where a person is able to exhale, research suggests that:

- people in a **positive** and perhaps **confident** frame of mind will predominantly exhale in an *upward* direction
- blowing in a *downward* fashion tends to be associated with **negativity**, **anger** or **frustration**.

In addition, the speed at which the exhaling takes place gives an idea as to the intensity of the emotion. So the faster the upward motion, the more heightened the feelings of confidence or positivity towards something. Conversely, the faster the downward motion, the stronger the adverse feelings.

Other idiosyncratic smoking gestures include extinguishing a cigarette. You've seen it a thousand times in movies, restaurants, meetings and during negotiations. (All of this pre smoking ban of course!) The way that a person puts out a cigarette reveals a lot.

- **The fast, intense drilling of the stub into the ashtray – mind made up** (ready for action).
- **The slow, careful, exaggerated gesture, by contrast – not quite sure** (something is troubling me).

Likewise, a person who inhales very deeply, in a **slow and deliberate** way, is usually in a pressurised state and the cigarette is more of a '**need**' than a displacement. You'll see smokers absent-mindedly tapping on the side of the ashtray, at a frequent rate (even though there's no ash on the cigarette). Similarly, they'll flick the bottom of the cigarette with an upward movement of the thumb to remove *non-existent* ash. This person's in an agitated state (whatever did you say to her?).

A word on cigars. Because of their size (and I'm talking about large cigars) you find different smoking gestures as well as different types of people compared to your average cigarette smoker (think Winston Churchill and boardroom bosses). Big generalisation here – because of the connotation of success that attaches itself to those people who wield cigars (rather like champagne), you'll find that many of these smokers (almost exclusively male) tend to use them as more of a 'prop'.

Cigars, unlike cigarettes, tend to go out if you don't take frequent puffs. So you're able to legitimately hold on to a cigar for a long time; a perfect displacement activity for the hand and a handy psychological crutch. As comedian George Burns famously said (he was 96 at the time) when asked about the cigar which he never lit but which just stayed in his hand the entire time during his stage and TV shows: 'At my age, if I don't hold on to something, I fall over!'

Self-comfort gestures

These are displacement activities that are **internally** directed.

Hand clasp

You'll observe people clasping their left and right hand together in what Desmond Morris called 'holding hands with himself' while talking. While it may seem like an innocuous gesture, just like everything to do with the hands it's usually telling us something, and it's not connected with confidence as it may appear. **The person is in a state of conflict.** The linked hands may move as the person speaks in a kind of rhythm – but all the time they remain locked because of the state of tension the person is in. The person is suppressing a negative attitude that they cannot display.

There are **two** main positions if a person is seated at a table or desk. **The elbows may be on the table with the clenched hands** *in front of the face*, or the hands *rest on the surface* of the table or desk. (Sometimes the hands may be a little lower and actually rest on the lap.)

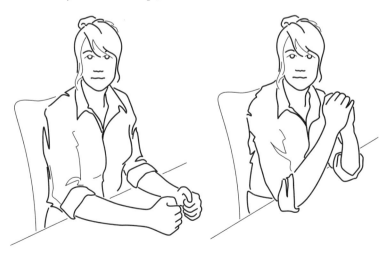

The interesting thing about these positions is that research shows that the *higher* the position (i.e. in front of the face) the more *negative* tension exists. This person would be resistant to you and your ideas. **If the hands are against or near the mouth area the person is holding back on a potential 'torrent'.** The person with the hands in the **lower** position would not be so agitated. They may be irritated at something, but they're not at a potential explosion point like the other person.

❝ This person would be resistant to you ❞

Sometimes you'll observe a person who's standing adopt the hand clasp with the hands positioned in front of *the thigh or crotch area*. This is usually quite different from the situations we've just illustrated. It's nothing to do with negativity or confidence – more likely nervousness. Many people adopt this **'crotch covering'** stance because they don't know what to do with their hands, and it's also a vulnerable position when you're standing front on so this offers a protection barrier.

'Steepling' of hands

It's a good time to talk about a hand gesture that seems to fascinate those of us who spend time studying non-verbal behaviour. It was first brought to attention by the anthropologist Ray Birdwhistell. He noticed that the type of person who didn't engage in animated body language during interactions – **in fact their gestures were, apart from facial, almost non-existent** – used the 'steeple'. What's that? It's a hand gesture that occurs in isolation of other gestures, so no 'clusters' here. *In fact it displays confidence.*

It looks very much like the pointed top of a church steeple and you may have seen it used by 'authority' figures when they're being interviewed by the media, or when they're interacting with other people (usually 'subordinate'). Apart from politicians, it's common to see professional people – hospital consultants, lawyers and the like – engaging in this action.

The gesture tends to come in **two** guises. When seated, if a person is speaking they tend to adopt the *raised* steeple. The elbows rest on the table, the two palms come together and the fingertips of each hand will gently touch each other and remain in that position – like hands that are praying. If a person is listening they may have the *lowered* steeple which can be resting on a surface or on or between their knees. It's usually a more cooperative gesture in this lowered position and the research also shows that women tend to adopt this lower steeple (in a listening role) rather than the raised one.

More men than women adopt this gesture and sometimes if, instead of an 'open' body language stance, the person tilts their head and body back they can come across as arrogant. Sometimes you'll even see people steeple above their heads. **All the studies show that when people observe steepling in others, they perceive it as a self-assured gesture.** Evidence also shows that people who switch from a clasped-hands defensive gesture to a deliberate steepling gesture experience completely different feelings.

TRY IT

Adopt the hand steeple in a conversation or meeting. See how it feels. Does it alter your emotional state? Does it make you feel more knowledgeable/confident? If you're feeling nervous and worried about your hands shaking in front of another person, try the steeple. It should steady you.

If you're a little bit nervous – say for an interview or an important meeting of some sort – the steeple prevents the slight tremor of shaking hands that is sometimes visible during these encounters. The more you do it, the more it becomes a habit, until it becomes natural.

If you find that you're unconsciously wringing your hands all the time, be wary of this and switch 'midstream' into the steeple position. Watch how your concentration improves as you find that instead of thinking about yourself, you're thinking about the other person.

There's another hand gesture that, unlike steepling, involves *interlaced fingers*. Normally interlaced figures are more associated with nervousness or defensiveness. When it's combined with a 'thumbs up' type of gesture, this hybrid display is a confident one. It's a positive gesture and to the onlookers it inspires confidence in the speaker.

❝ *It inspires confidence in the speaker* **❞**

Hands behind the back

You can no doubt recall someone from your past, or someone you currently deal with, who puts their hands behind their back and grips them together. It's not a hiding of the hands gesture – **on the contrary it projects confidence** because you're happy to expose the vulnerable front of your body. We've all seen policemen, higher ranking military, hospital surgeons and university lecturers adopt this stance.

It's been shown that adopting this stance can actually give you more confidence and change your feelings and the impression that people have of you. I remember many years ago showing three visitors around some advertising agency offices. At the end of the tour I instinctively handed over some annual reports to the person who had been walking around with his hands behind his back. It turned out he was an IT trainee (he'd been there six weeks) and the other two were the managing director and financial director.

Arms

There's a wealth of **difference** in what a few inches can convey about an attitude. If the second hand behind the back is not gripping the other hand but the *wrist*, we're talking about a situation in which the person is annoyed or frustrated at something. (It can also signify nervousness.) **The hand is almost controlling the wrist to prevent it being unleashed.** The further up the arm the hand grips, the more anger or frustration the person is experiencing.

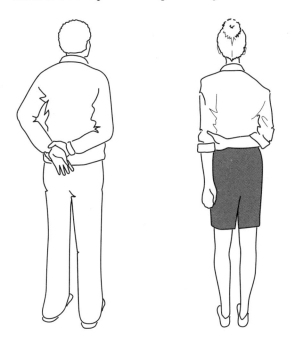

Hand to face/head

When under stress or discomfort of any sort, we'll frequently engage in self-comfort gestures that involve the hands touching the face, the head or the hair. (Does this ring a bell with your own behaviour?) The interesting thing about these behaviours is that we tend to *touch the areas that may have been used to comfort us as children.* So the school of thought is that we're comforting ourselves in the way that we were consoled in our childhood.

For example, if we're undergoing any kind of anxiety we might *brush our hair with the hand* (**top of the head or back of the neck area for men, typically**). Does that sound familiar to you? **Ever watched a 'celebrity' on a chat show after they've just sat down and the applause of the studio audience has subsided?** Up goes the hand to smooth the hair on the top or back of the head (usually male). **Or the person returning back to their parking space and seeing in the distance the sight of a traffic warden printing out a 'ticket'?** Up goes the hand . . .

David Beckham

Another variation of this – and quite often you'll see it with both hands – is the sight of David Beckham or some other footballer as they miss the goal in a penalty shoot-out, or a tennis player as he or she double faults on match point at Wimbledon. The player's hands will instinctively go to the head or back of the neck in a self-comforting gesture. The point is, it betrays some kind of displeasure, anxiety or momentary nervousness. Quite evident what it means in this case – millions of 'gutted' fans would probably be able to interpret this element of body language without any trouble. This *'cradle'* gesture with both hands can be seen in any situation in which we respond to bad news or anxiety and need protection. Memories of parents supporting the head as a baby provide the self-comfort, which – for that fleeting moment at least – provides us with security.

Our skin has *receptor cells* which respond instantly – because of their sensitivity – to touch. They send messages to the brain and provide us with comfort. Never underestimate the power of touch (a little more about that in Lesson 5). So we frequently resort to self-comfort actions that almost recreate (sorry, taking you back to childhood again) how our parents and other adults made us feel secure when we were babies or toddlers.

Interpreting **hand-to-face** or **hand-to-head** gestures requires a little work, initially by becoming aware of how people typically exhibit these and at what point in a conversation they may frequently occur. This provides us with clues. Of course you've built up a reservoir of knowledge over the years, which you instinctively tap into. But, as with all the other analysis that you'll be doing on a day-to-day basis to become proficient, you need to pay attention and become super-observant.

❝ You need to pay attention and become super-observant ❞

Boredom

Quite often you'll notice the hand-to-face gesture that gives a clue that the person listening to you is, for whatever reason (that's for you to deduce!), not exactly riveted to your words of wisdom. (Equally you may display this when you're on the receiving end.) This may occur during a work presentation, with friends in a coffee bar, at a party or gathering, during a training session – it happens everywhere.

The typical hand-to-face gesture is one where *the hand supports the entire jaw with the elbow resting on the table*. Generally, the more you see of the hand covering the jaw (the entire cheek), the more bored or disinterested the person is. So if the whole hand covers the entire jaw right up to the eyebrow – sounds like serious boredom. When we become disinterested or bored with another person it gradually builds up momentum, to that final hand position which is holding up the head.

CAUTION

Sometimes, depending on the positioning of the hand on the jaw, it may just be a switch to a position that intensifies concentration by the listener, either because it's something that concerns them or because they want to display empathy to the listener ('Oh that's awful. He really said that to you? . . .'). In other words – the opposite of boredom. (*See why body language is so confusing?*)

If a person was showing attentive signals prior to that and then suddenly switches to this extreme position, it's unlikely to be a boredom gesture. As we said earlier, boredom or disinterest usually builds up momentum and doesn't switch that easily.

Sometimes you'll see *the hand folded in a fist* arrangement with the fist gently holding the lower part of the cheek. It may be a boredom signal but sometimes a person switches to that position when hearing something disturbing, or something that 'touches a nerve'. Later, the hand may be removed from the face back to its previous resting position. It's the eyes that will *confirm* what's being felt – the cluster, remember?

However, *the fist supporting the chin* is a different matter – just as with the entire open hand covering the whole of the cheek or jaw, they're both propping up the head. So, it's a danger signal if you're giving a work presentation, training session or speech and you see a multitude of heads being supported in this way.

If they're accompanied by **doodling on pads, daydreaming** (look at the eyes!), **blank expressions, a fake smile, the occasional sigh, tension in the jaw** (we're now getting some cluster activity) and **increasing movement of limbs** – do something!

To check for the boredom signal you need to *look at the eyes* in conjunction with this hand-to-face gesture. If the eyes are looking down, or half shut, or there's a bleary-eyed look (that almost says 'please stop the fight'), then you can almost be sure that you need to change tack to try to re-engage, or decide that the situation is hopeless and abandon your cause.

Disapproval or disagreement?

Sometimes you'll see a person perform a mouth cover, whereby *the fingers of the hand cover the mouth with the thumb pressing against the cheek*. Usually it indicates that the person disagrees with what the speaker has said or doesn't believe they are telling the truth. **The hand subconsciously goes to the mouth area.** The interpretation of this gesture? It's as though the brain is directing us to block ourselves from blurting out the fact that we disagree, disapprove or distrust what the person is saying. Take heed if you see this action in an audience of yours – be it displayed by one or many listeners. Again, as we've said many times before, change tack. Try to get to the root of the problem and look for other clues.

When 'the boot is on the other foot', and *the person who is speaking* performs this gesture immediately after saying something, **it may indicate that they're lying about something**. Consequently, the neurological instruction (from the brain) is to try almost to prevent more untruths from coming out of the mouth (we'll look at this in Lesson 6).

A variant of this is when you see somebody put *their fingers in their mouth*. This usually denotes that the person is experiencing discomfort or pressure of some kind. Any kind of anxious situation makes us want to put something in the mouth. Just being put on the spot to make some kind of everyday decision can produce this action. I've seen people staring at railway timetables adopting this pose as they anxiously work out connecting train services. This comfort gesture takes us back to childhood, it is said, when as young children we looked to the mouth to placate us, from feeding to thumb-sucking.

❝ *This comfort gesture takes us back to childhood* **❞**

So, if you find yourself doing this when you're not even conversing with anybody, and may even be on your own, *check your own feelings.* **The whole point about body language is that it not only tells other people how we might be feeling but also provides us with a reminder of our own internal dialogue** – which has created the feeling. *We know that if we change the thought then we change the feeling* – which changes the non-verbal behaviour.

So if you're with somebody be aware that this gesture may tell them something about your state, which may not be what you want to broadcast. Equally if you see this happening with the other person you have an opportunity to delve into the circumstances and to see if it is anything related to you. If it's not, you at least have had the chance to show some empathy through your questioning.

BODY WISE

The way you feel affects the way your audience feels.

Another common hand-to-face gesture is one where the thumb supports the chin and the all-important index finger points in a vertical direction up the side of your cheek. **It's not generally a gesture that denotes positive feelings but rather disagreement of some sort**. There's often confusion with this pose because it is similar to one denoting evaluation (rather than negativity), **but it's the positioning of the thumb** that provides us with the clue. You'll often find that the person holding this position will occasionally move the index finger a fraction, to touch the eye (maybe a rubbing motion), blocking you out even more!

We'll often see the eye rub in isolation to any cheek or chin touching. Sometimes it involves pulling down an eyelid.

The most common actions

It was noted in a survey of hand-to-head/face actions in **stressed** situations that the most common were (in decreasing order):

1 jaw support
2 chin support
3 hair touching
4 cheek support
5 mouth touch
6 temple support.

The study showed that these actions were common in **both** males and females. Temple support was more common in men by a 2:1 ratio; with hair touching there was a 3:1 bias in favour of women.

BODY WISE

Our self-comfort gestures of stroking, hugging and touching, when experiencing discomfort, are said to relate to childhood because they are the subconscious gestures of another person's consoling touch.

When someone switches to positive 'evaluation', we're entering into more favourable territory. There's more of *the chin* and less of the rest of the face that our hands gravitate to. A person who may be interested in your message often adopts *a lightly closed fist resting on the cheek with the index finger pointing upwards, usually by the ear*.

The next stage, when you ask for interaction or for a person to make a decision about something, is usually a switch of the hand to the chin. If it remains there, or perhaps they lean forward and remove the hand from the face altogether, they're still interested in your presentation, speech, proposal, training session, holiday snaps, recipe for blueberry muffins, pontification on falling house prices or the rising price of fuel. If they switch to a cluster of boredom signals – there's more work to do.

Arms

You remember we spoke about open body language and closed body language? By now I'm sure you've taken in (something that you instinctively knew) the fact that it's only when we're in an 'open' position that things run smoothly. Well, it's no different with arms. We'll be talking about arm positions and what they signify, and of course what effect it has on your own brain when your arms are fixed in a certain position.

❝ When we're in an 'open' position things run smoothly ❞

Folded

Arms are particularly interesting in body language terms because in a folded position they have the capacity to form a protective 'barrier' between people and close down communication. Have you seen 'bouncers' standing outside Fifi's nightclub at 1 a.m. – can you remember what their arms were doing? What about inside the nightclub (you managed to get in?)? What were the girls doing to show you that they weren't interested?

Since we seem to have to revert back to **childhood** to find the answers to much of our adult body language action, think back to how you used to hide behind things for protection as a child. If it wasn't a parent, it might have been a tree or table or anything that provided a defence between you and the outside world.

As we grew up we discovered we had our own *portable* defence mechanism – our arms. They could be crossed – in a number of ways – in times of discomfort. **We adopt defensive postures** when we hear something we disagree with, if we're in a tense situation or if a negative thought passes through our mind creating a corresponding feeling.

BODY WISE

Crossed arms (aside from temperature reasons) nearly always signifies some form of discomfort.

The crossing of arms is a gesture that you see throughout the world. As mentioned earlier, it is essentially signifying discomfort on your part, and as such is transmitting that to observers. We're always worried about what our body language signals are conveying to other people, right?

Of course, before we start jumping to conclusions about people we should note that there are some people who have a habit of crossing their arms – in a loose manner – as a 'self-hug' comfort gesture. It almost becomes their 'default' position to show they're relaxed and ready to hear your conversation.

A good example of this is the character Rachel (Jennifer Aniston) in *Friends*. She folds her arms to get comfortable to listen to the latest tales of woe from Monica, Joey, Ross, Phoebe or Chandler.

Jennifer Aniston in her typical, relaxed pose

Whenever we humans experience anxiety or distress we tend to withdraw our arms. The arms will always go straight to the sides or will close across the chest. This provides us with a defence barrier and – just as important – self-comfort. Since we're always concerned not only with what signals we're picking up but also with what we are giving out, we should be careful when adopting any crossed-arm position – even if the central heating needs turning up!

BODY WISE

Studies confirm that people witnessing crossed arms in another person will *interpret* it as a defensive or negative gesture.

There are a few crossed-arm positions that you no doubt come across in personal and working life. The common one, which is probably the one you recognise in yourself from the time when your genetic 'hard-wiring' first introduced it to you, is *the general crossed arms*. Both your arms are folded across your chest (one or both hands may be tucked under between the arm and chest). **This is the most common arm cross.** It's seen throughout the world and usually silently screams out the same thing (oh so politely): 'I'm feeling defensive or negative about some*one* or some*thing*.' As well as in your interactions with other people, you may see this gesture as you go about your daily life: on tube trains, in lifts, long post office queues, waiting rooms, at social functions. Anywhere that people are experiencing discomfort or insecurity.

At work, during a meeting or a discussion that you're having with a colleague, client or boss, for example, if it suddenly occurs (when 'baseline' behaviour tells you it's **not** normal) then you may have come up with something **that they don't**

agree with. They may be verbally indicating that everything's okay (this time with a hesitant and softer voice), but this gesture, perhaps accompanied by a change of facial expression (less 'open'), should set alarm bells off.

" This gesture should set alarm bells off "

Consider the following scenario:

Simon (in departmental meeting) talking to his PA: 'Okay, let's firm up on that murder mystery weekend in Brighton. I'll fix that four-star hotel, the one on the seafront. You know – the one with that atrium – it's got that health and leisure club in the basement? Bring your boyfriend – my wife's coming. We'd like to meet him. (Karen's hand covers her mouth.) What's it called . . .?'

Karen: 'Yes, Thistle Brighton. It is a lovely hotel, very atmospheric.' (Her voice pitch and tempo is different and her words tail off as her hand moves down to clasp her neck.)

Simon: 'That's it. Great view of the sea from the restaurant. Handy location for the shopping too.

(Karen's now looking downwards and has also crossed her arms.)

Well, Simon didn't pick up the signals. All the obvious clues were there – there was definitely something wrong. He should have known that her 'baseline' behaviour included **good eye contact** at all times. That should have been the give-away that something was wrong. Then there was her speech – **hesitant**; she **covered her mouth, clasped her neck** and went into an **arm-fold. A 'cluster' of anxious behaviour**. It turned out that she had just split with her boyfriend that weekend.

So there was a 'cluster' informing us that the genuine message was leaking out from the non-verbal actions. Since we know that the true meaning (because it's hard to fake) of any message is from the body, what we observe from the body and vocal signals (the 55 per cent and 38 per cent) tells us that we're not going to get anywhere with this person unless we try to find out what's wrong.

Certainly, as long as the person remains with this defensive arm-fold barrier their attitude is not going to change. '*Something's gotta give*', as the saying goes.

When I'm confronted with this situation, I usually try to get the person to *do* or *look at* something. If you hand them something to look at or get them to do something, it makes them break the 'deadlock' of their arms.

❝ *Try to get the person to* do *or* look at *something* ❞

I remember when I was giving a presentation at work with a female colleague and she sensed that the client had suddenly turned into '*not interested anymore*' mode. There was no legitimate or relevant document or brochure to hand him. She picked up the new jar of coffee and pretended it was too tight for her to open and asked the 'now beaming' client if he would oblige. Arms unfolded now, a premature coffee break and also an informal interlude to ask if there were 'any points that needed clarifying'. (Never underestimate the power of a *damsel in distress*!)

Types of arm-cross

The important thing about the arms-folded gesture is that the demeanour and mental attitude of the person is *fixed* in that mode as long as they *continue* with that posture.

Arms gripped

Another type of arm-cross that occurs is where *the arms are gripped*. **The hands grip both of the upper arms in a tight fashion as if to ensure that they keep the arms down, so that they won't get away.** It's like the tight gripping of the arms of a chair. The person may be experiencing extreme anxiety or anticipating an unpleasant event, or feeling extremely stubborn about something. You see this pose during union pay negotiations with the dissatisfied members listening to speeches by opposing bodies. If you're displaying this in front of other people, you can imagine the image you're conveying.

Crossed arms (with fists)

At other times you may see a *crossed-arm position with fists*. Doesn't sound good, does it? The fists are placed under the arms. Not only have we got defensiveness here, there's potential hostility on the menu too. Facial expressions may confirm these feelings, as well as perhaps foot or leg movements. Make sure you don't 'mirror' this pose if you're trying to achieve a conciliatory outcome. **Use your open body language to encourage this person to open up and disclose the reason for their feelings.** You know what the feelings are – you want to know *why*. You may see nightclub bouncers, security personnel and policemen adopting this pose. It gives out the signal they deliberately want to give out.

Crossed arms (with thumbs up)

There's another arm-cross that body language researchers have come across more and more. It involves *the thumbs* in an upward position. I've seen it a lot with tradesmen when they're negotiating or justifying their prices to householders faced with some emergency repair, or with car mechanics in the same scenario.

The crossed-arms stance conveys discomfort or apprehension, but the **thumbs up** conveys a certain amount of **confidence**.

There's something about the thumb – what is it? We know, for example, that confident people often have a thumb (or two) sticking out of a jacket pocket. The late John F. Kennedy – looking at all the old newsreels – always seemed to have his hands in his jacket pockets with the thumbs sticking out.

Yes, the builder and car mechanic we spoke of earlier feel uneasy about the process of maximising the amount of cash they would like to extract from you. But at the same time they feel confident because they know you're faced with a particular situation: you're in a fix (and they know a man who can get you out of it!).

So you'll often see this type of arm-cross both in and out of work, and where it is in evidence in a working environment you know that your colleague or the person you're in charge of, while experiencing *unease* at the time, also feels *reassured* about their position.

The partial arm-cross

Before we leave this important topic we'll take a look at the way we may try to minimise the display of this defensive gesture. You probably do this yourself from time to time. You'll engage in a *partial arm-cross* – **with just the one arm**. Just the one arm will form a defensive barrier by reaching across to grip the other arm. It's quite common with women and you'll see it a lot when groups of people are conversing, especially if they don't know each other well.

Back to childhood for a moment – it reminds us of when we were held by the arm or hand when we were afraid. You'll see this single arm-cross as candidates are listening to results on the stage on election night in their constituencies. Also, it's very evident when people are waiting to receive an award or prize for some sort of contest.

Sometimes, this partial arm-cross shows a variation. On occasions we're not in a position to grip the other arm. We don't have the opportunity for self-comfort so we have to resort to **disguise**. We don't want people to know how we're feeling. **But we've still got to displace nervous energy and we still need the comfort of a barrier, however fleeting.**

❝ We still need the comfort of a barrier ❞

It won't surprise you to know that people in the public eye, like politicians and 'celebrities', need to keep up an image and therefore engage in *disguised* defensive gestures all the time. We're privileged to put the microscope on them usually through the TV camera. So they have to work extra hard to 'look the part' even though inwardly they are experiencing great discomfort.

Glossy magazines are always picturing couples arriving and leaving functions and nightclubs. Who's she with? Who's leaving with him? How is she holding him – looking at him? What's the body language telling us? The slightest nuance of any revealing body language will be taken down, Ms (Kate) Moss, and used in evidence against you.

So celebrities and people in the public eye (who have an image to project – and protect) may typically have an arm crossing over their body if they're walking on a red carpet, or entering Downing Street, or walking up the steps of a fashionable restaurant in LA. An arm may cross over the body to straighten a watchstrap (sometimes there is *no* watchstrap); another person may straighten an already straight tie. An arm will cross over J.K. Rowling's body while walking on the red carpet and her hand will rise to her ear to straighten an ear-ring. **All of these things are handy defence mechanisms in the absence of a permanent barrier.**

BODY WISE

You are what you project. Actors, politicians and other 'celebrities' are trained to be acutely aware of self-presentation. The manner in which we project ourselves is an indicator of our mood as well as the level of comfort or discomfort we are feeling.

When it comes to our arms, if we're in a relaxed state, the arms are by our side. It conveys an impression of openness and confidence (even if you're not feeling that way).

HRH (or hands revealing hesitancy!)

In England, the Royal Family are often cited when we're talking about disguised gestures, purely because of the sheer number of engagements that they have to attend – the opening of this, the opening of that – and so we constantly witness the obvious discomfort that they still feel, even after years of practice.

HRH The Queen is noted for the handbag that is constantly over her arm in front of her body, giving her a barrier at all times when she needs it. Prince Charles never disappoints – he has some 'endearing' gestures, especially the 'signature' one that he uses to reduce anxiety. Typically, as he leaves the royal car and walks past the crowds towards, say, the Royal Albert Hall, his

arm will cross over to tamper with his cuff-link. Sometimes, he'll treat his audience to a second 'signature' displacement activity. Even though Royals are known for not carrying money, he'll cross his arm over and pat his jacket around the inside pocket area to make sure his (non-existent) wallet is still there. **A wallet no doubt containing money that he would need for purchasing an ice cream during the interval and an 'Oyster' card for getting back home to the Palace after the show.**

Hand to neck

The hand instinctively goes to the neck in many instances when we're undergoing anxiety. Women tend to touch the neck more often than men. As a self-comfort gesture we'll stroke the front or – commonly with men – massage the back of the neck. Men will typically enfold the area right under the chin with their hand, stimulating nerve endings.

It's noticeable that women experiencing anxiety of some sort will stroke the *front of the neck* at times, or if they're wearing a locket or necklace will fiddle around with this. When emotions are running higher it's interesting that women touch or completely cover the neck 'dimple' area just above the breastbone. **They may cover the area with their hand the whole time until the feeling of 'distress' has passed.**

I noticed an interesting display recently in a hotel lounge with a widescreen TV. A number of guests – men and women – were watching one of the later episodes of *Masterchef* on the BBC. The tension was high as the nervous contestants stood there waiting for the results to see who would be leaving the show that week – their autonomic nervous system displaying their heavy breathing. I noticed **four of the seven** women seated around the TV bring a *hand up to the neck* to clamp it. The hands stayed there the whole time until the result was announced.

So, when you're observing body language cues from women take a look to see if the hand goes to the neck and what the position is. From handling a necklace, to stroking, to a full clamp – this should give you an idea as to the intensity of any discomfort. Again, look for the clusters in order to make a more-informed judgement.

❝ Again, look for the clusters ❞

Hand to eyes

We looked at eye 'cut-offs' in an earlier Lesson. We'll bring the hand as a shield over the eyes when we're faced with seeing something unpleasant, or someone has said something to us that provokes an instant feeling of stress and anxiety. The hand is shielding us from 'seeing' the reality of life – even if we're about to *hear* something unpalatable.

Ex-PM Gordon Brown's discomfort is relayed around the world

An illuminating – and painful – example of this was provided to viewers during the 2010 election campaign in the so-called Duffygate scandal. The campaigning prime minister Gordon Brown was in the BBC studios on the *Jeremy Vine Show* when the host told him that reporters had picked up some of his private comments in his car, after he'd spoken to a constituent – because he'd left his microphone on. The PM began to apologise for what he'd said because they were private comments, but then he was told, to his surprise, that a recording of it was just about to be put on air. Gordon Brown's eyes closed, his hand went to shield his eyes and his head slumped down almost to the table. The recording was then played. This shot in the studios was shown to TV viewers around the world.

It's a measure of the impact of this interview – and the visual images – that it won the 'Interview of the Year' award in the prestigious Sony Radio Awards held in London in May 2011.

Feet

Like the arms, we can pick up a lot of information from leg gestures, especially when they are accompanied by arm signals to form an all-important cluster. Feet are important too – we're conditioned to turn *towards* things and people who we like. **Turning the feet away shows that we want to distance ourselves from a situation.** Usually, we're unaware of which way our feet are facing – until we actually look down.

As you'll see in Lesson 5, it's this lower half of the body that is most revealing in the detection of lies. What we call *intention movements* are very revealing in terms of legs and feet. For example, if you're conversing with someone facing you, and she later on in the conversation turns one of her feet in an outward direction – the other one still facing you – even though you're getting on fine, it's a signal she needs to or wants to go. **Her body is still facing you front on, but one foot is pointing elsewhere** (towards the exit point). She may have another appointment to keep or has decided that she just wants to go. The feet provide a great deal of information about a person's attitude.

Picking up an intention cue quite often saves a relationship because some people, out of politeness, don't want to verbalise what their intentions are. They rely on the body language skills of the other person – that means *you*. If they have to keep making repeated intention movements, it can irritate. Often it's the last thing they remember about the interaction and so it can cause negative feelings.

Sometimes there's no confusion. You're talking to someone – they're standing to the side of you with both feet pointing away from you. Despite the pleasant facial expression, **it's obvious from their feet as to where they're heading.**

The reason why the feet are regarded as being one of the most 'honest' parts of the body is, according to some researchers, that since the beginning of time the feet and legs have had to react to threat and danger *without conscious thought*.

Our lower limbs give out information about negative and positive thoughts. You may bounce your feet up and down with excitement while you're sitting in the concert hall waiting for Madonna to arrive on stage. Equally your crossed leg may be kicking a foot up and down impatiently in a reception area because the interviewer is running 30 minutes late for your job interview.

Towards or away?

As we briefly touched on earlier, it's in the nature of human beings **to turn towards what pleases us** – whether it's things or people. We can observe this in standing behaviour and sitting behaviour. If you're standing and talking to someone it may begin with your feet not pointing towards the person – in other words, only from the hips up are you facing the person.

As the conversation progresses, you or the other person may gradually, without noticing, move the feet to face the other. It may happen quite naturally. If the feet stay pointed away (and towards the exit) **then you don't want to stay, or rapport hasn't been established** for whatever reason. Sometimes the feet may shift away during the interaction – it may be because a person has to go, or they are feeling discomfort and need to leave. You need to know why this has happened – a pressing engagement or disapproval?

During seating it's often helpful – assuming that legs aren't crossed – **to note where the feet are pointing**. If both feet are on the floor and one or both are not facing you, *you haven't quite attained that stage of interaction whereby everybody's comfortable yet.* Be aware of whether it's due to seating position or a person's state of mind.

You may equally have a situation where a person is sitting **cross-legged** in a chair with their **feet facing away from you**. Their crossed leg acts as a barrier against you. Again, check whether it's due to their emotional state or the placing of furniture. Also:

- Was it a change of sitting posture?
- Was he previously **facing you** and has just now moved into a position **away from you**? (Was it something you said that didn't go down well?)

❝ *Check whether it's due to their emotional state* **❞**

At the next opportunity take a look at a group of people standing and talking. Do any of them have one foot pointing at a *particular person* – for example, a male with one foot pointing towards the female in the group, or vice versa. Watch for this on social occasions and at business ones too. It can reveal a lot (about you as well). **We don't usually know what our feet are doing.**

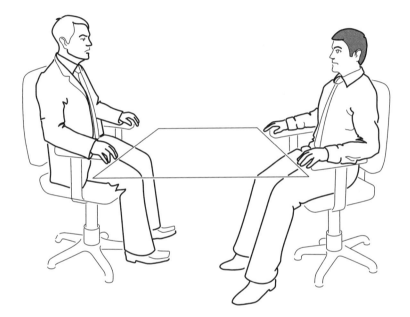

Intention gestures

A common gesture that we all engage in is when we're in a sitting position and we want to signal that we want to leave. Quite often it's subconscious. We'll adjust our feet to point to the exit. The hands would go on the armrests (or on the knees if there aren't any) and we attempt to leave. If the person we're with fails to pick up on the signal and carries on talking, we have to go through it again – and again. *It can cause irritation if the signals are not picked up.*

BODY WISE

Being the furthest point from the brain, we're usually unaware of what our feet are doing and where they're facing.

Apart from pointing in a certain direction, if we notice – either when standing or sitting – that a person's feet are fidgeting, this displacement of energy tells us that the person **can't wait to get out of the situation**. Similarly, tapping of the feet (as with fingers) denotes a person running out of patience.

Legs

Just as feet can provide a clue as to what our intentions are or *where we'd really like to be,* so the legs can reveal a lot, especially as part of a cluster. Leg crossing (which is usually right over left) often forms part of a cluster with the arms to reveal negative or defensive attitudes. **Unlike crossed arms, we can't take the leg-cross in isolation** as an indicator of *discomfort.*

BODY WISE

Be careful in interpreting crossed-leg gestures with women because it may be for comfort reasons or because of clothing – as opposed to being *attitudinal.*

It is with the crossed-arms position then that the crossed legs reveal anything, because even with men it's sometimes adopted out of habit or because of an uncomfortable chair. **Generally, for both men and women, crossed arms *and* legs suggests trouble.** As we noted earlier, something has to be done to 'open them up', and at the very least to find out what's causing the negative attitude.

Sometimes it's the *ankles* that are crossed rather than the legs. With a woman the knees tend to be held together – hands may rest on the lap. With a man the legs are usually spread out and the hands may rest on the sides of the chair. Again, as we've said before, it's the impression that it relays to other people that's important. If an ankle lock makes you look defensive – and therefore not 'open' – is that what you really want to convey?

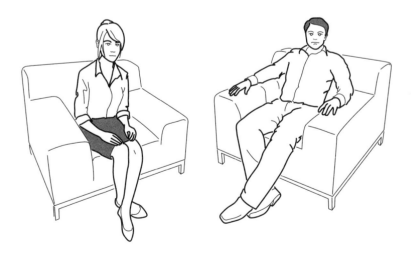

Just a word about an irritating 'crossed' position that is exhibited by men (often referred to as the 'four-square' position). You've seen it before and at times (just you males) you've been guilty of it. The 'superior' hands behind the head and one leg crossed over with its ankle supported by the knee of the other leg. The *'one day you could be as wonderful as me'* pose. There are certain professional people who exhibit this with their subordinates. Women are particularly irritated if they're in the same room and they see it, even if it's not directed at them. It may be okay for a Hollywood film producer but away from the glitz it's a turn-off.

BODY WISE

The feet and legs are the most 'honest' parts of the body.

Proxemics

The American anthropologist Edward Hall introduced us in the 1960s to the concept of 'proxemics', deriving from proximity – nearness in other words. He spoke of 'personal space' and how as territorial animals we impose an invisible 'bubble' around us for protection.

We do it at work by forming **barriers** in our offices, for example. The fashion for open-plan offices leads to staff using things like framed photographs, oversized staplers and pot plants to erect barriers between them. If you travel on trains and the London Underground (that's a double-whammy) you're probably used to *'space invaders'* while you're standing gripping tightly onto your two inches of clear rail. Your normal 'bubble' has been pricked and the only way you can cope is by:

- **no eye contact**
- **turning the head away**
- **closed body language (as you pull back) – and maybe a newspaper as a barrier.**

Somehow, as psychologist Robert Sommer suggested, we're able to cope with uncomfortable crowded situations and a *violation* of our personal space by tricking our minds into believing that the person is inanimate. As such, the normal conventions of social behaviour don't apply – if they accidentally touch us we ignore it, we adopt a blank expression and we also avert our gaze. A similar level of behaviour pattern exists in lifts as they go up and down floors.

In normal everyday interactions, taking account of humans' spatial needs, Hall suggested that there were **invisible** *zones* that were suitable for specific situations. **Depending on the relationship you had with somebody, each zone represented how close you allowed that person to come near you.** Mmm . . that's interesting, isn't it? How many times have you been turned off ladies (question specifically for females) by someone who comes just a little too close to you for comfort? In a 'stranger' situation, social situation, work situation or flirting situation – maybe all of these?

These are people who just don't quite get it, do they? Either they're taking liberties, or they plainly don't know (never been taught) – or maybe they're from a different culture. Whatever – it's irritating, isn't it? It's not just confined to females of course. We all experience it, male and female alike.

How many times have you 'rejected' a person because their distance just didn't feel quite right? Everything else was okay, but they didn't respect your personal space 'bubble' – they didn't know what the appropriate 'zone' was.

Appropriate zone data

- **Close intimate:** 0–6 inches. Reserved for very few – lovers or people (e.g. your children) who you don't mind touching you (as such, the most intimate behaviours are permissible).
- **Intimate:** 6–18 inches. Only the most important people – lovers, close relatives, close friends – are permitted. If strangers or people you don't know well, or don't like, violate this space, you feel uncomfortable.
- **Personal:** 18 inches–4 feet. Arm's length – you are able to shake hands in this zone. This is the ideal one in the western world for most personal interaction. You'll notice this distance at social functions, maybe office functions and parties, or 'networking' events. (Sommer noted that if you stand *outside* this space, it can cause negative feelings in the other person.)

- **Social:** 4 feet–12 feet. People who are not familiar to us who we may have to interact with – for example, tradesmen or shop floor assistants in stores.

- **Public:** 12 feet and over. When addressing a group of people in a formal setting, this was considered to be an acceptable distance from the front row. Usually there's no social interaction in this zone.

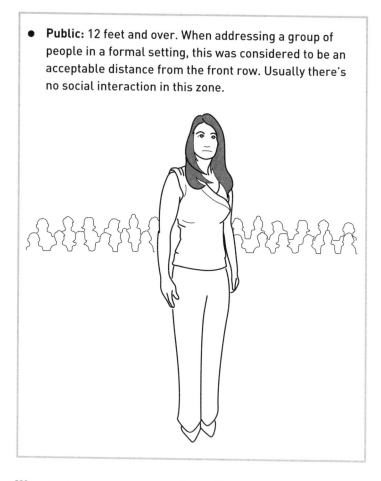

We can gauge, or more realistically guess, the extent of a 'relationship' between one person and another by the distance between them – and also, at times, how they feel about each other.

BODYtalk

Q Should we be striving for 'open' body language most of the time?

With your normal relationships, yes. If you came across a grizzly bear nobody would be surprised if your posture and limbs adopted a more closed position and formed some kind of barrier and adopted a position of defensiveness. So, generally in most of our interactions in life, we'll assume that we're looking for harmony and cooperation, and it's open body language that is ideal for achieving this.

Q I'm worried about my arms now. I didn't know that for years I've been crossing them in various ways. When I'm with friends, during meetings, at the theatre, standing in line at the bank – I suppose if I analyse it, in most of these situations I'm not feeling great. So according to what you told us earlier, I'm reflecting my feelings to the outside world. No harm done in most cases I suppose, but I guess my friends could be offended while they're telling me that long-winded story; and also in meetings, I'm wondering what messages I must have been giving out.

You're right. You may get absolution for some of these instances. But remember what we said – it's what the other person perceives that's important. What you need to be aware of from now and always is that the longer you stay in a particular cluster gesture – let's say folded arms with folded legs – the longer you keep that feeling. So you get to the stage where first of all the mind controls the body. Then, when you've adopted a certain position, the body controls the mind. So effectively you're stuck.

So the way to get out of this rut has to be, initially, through self-talk. Since your thoughts control your emotions, which then tell your limbs how to behave, it's only your new thoughts that can get you to

unlock. Unless you're lucky enough, in this instance, that your girlfriend is in the theatre with you and hands you a bar of chocolate.

Q Can we talk about these displacement gestures? I now know that I'm at it all the time. I'd like to dazzle people with my knowledge the next time we're having a drink. I missed the definition you gave earlier. Can you give me a one-liner?

Sure. It's any gesture we make to displace nervous energy.

Q Can I twist your arm for a similar one-liner for self-comfort gestures?

Well – I'd rather you left my limbs out of it, but here goes. It's a gesture directed towards your own body to comfort you. (Much of it goes back to childhood.)

Q I'd like to ask about these hand-to-face activities. So, if I see someone touching their nose when I ask them if there's any problem with the roof – we're going for a second viewing of a house in a week's time – can I assume they're lying?

No – ID 10T error. We'll be discussing the nose touch in Lesson 5. They might have an itchy nose – just at the time you asked them that question. However, if you noticed a pattern in this activity every time you asked an awkward question, and also if there's a cluster of other signals, you could be barking up the right tree.

Q I find myself gripping the arms of a chair and putting my feet under the chair when I'm with people. They'll understand that it's not a defensive or nervous gesture, won't they?

No.

Q I find myself always gripping my hands with fingers together when I'm summoned to a meeting at work. Is my subconscious telling me something?

Yes – that you're experiencing discomfort. And it's telling the astute members of the audience too.

Coffee break . . .

 Of all our body parts the hands are the most spontaneous.

 There are more nerve connections between the brain and hands than any other part of the body.

 An 'open' hand in conjunction with the palm up position is associated with honesty and trust.

 Things you do to help displace anxiety or nervous energy are known as displacement activities.

 Some displacement is externally directed, such as fiddling with keys or a pen.

 Self-comfort gestures are internally directed.

 With the hand clasp where there are usually two main positions, the higher the position, the more nervous tension exists.

 When the arm behind the back grips the wrist it usually signals annoyance or frustration.

 With hand-to-face or -head gestures we tend to touch areas that may have been used to comfort us as children.

 Never underestimate the power of touch because our skin has receptor cells that are very sensitive and responsive to the power of touch.

The typical hand-to-face gesture for boredom (while seated) is with the hand supporting the entire jaw with elbows resting on the table.

 Always look for a cluster of gestures to support your assessment.

 To check for boredom signals you need to look at the eyes.

 When the listener engages in a mouth cover with the fingers covering the mouth, it usually indicates they disagree with something or that the person is not telling the truth.

 If a speaker performs a mouth cover directly after saying something, it usually indicates they are lying.

 If a person has their fingers in their mouth it usually signifies some form of deceit.

 When the thumb supports the chin with the index finger vertical it is not usually a positive sign.

 With the evaluation gesture it's the position of the thumb that gives you the clue.

 In body language terms the folding of arms has the capacity to form a barrier.

 Studies confirm that people seeing crossed arms will nearly always interpret it as a defensive or negative gesture.

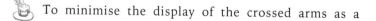 To minimise the display of the crossed arms as a

defensive gesture you'll sometimes see the partial arm-cross (quite common with women). Also, especially popular with celebrities and people in the public eye is a variation of this, the disguised arm-cross.

 Unlike crossed arms, we can't assume that crossed legs are a sign of defensiveness; we have to look for a cluster of behaviour.

 Feet provide a good clue to a person's feelings and intentions – note whether they're pointing towards or away from you.

 The study of personal space is known as proxemics – there are four identified zones.

 The ideal zone in the western world for most personal interaction is the 'social', which extends from about 18 inches to 4 feet.

Lesson

5

'Perception is real. Even when it is not reality.'

Edward de Bono

Lying

Advertising agency executive Charles, returning from a client meeting (speaking to his boss):

> *Charles*: 'Great news. I managed to convince them that they should take the back cover for the next available issue.'
>
> *Boss*: July?
>
> *Charles*: 'No, I didn't have to. They convinced themselves.'

The area of lying and deception holds great fascination for many people and is the subject of much scientific study with psychologists and other researchers continually exploring the area.

The effectiveness and success of any interaction is determined by how the listener *perceives* the message (and you!), not by what you may have *intended* it to communicate. So it's perception – on the other person's part.

❝ Deception *is often just* perception **❞**

They'll base their perception on your body language behaviour as well as your words. But, as we know, we can all give off negative signals that may not reflect us in the right light. We may give the impression of deceit when it's actually not the case. Equally we may interpret another person's body language signals as them not being entirely honest – and that may not be the case. *Deception* is often just *perception*.

We're discussing more self-serving types of lies in this Lesson, rather than the social type referred to above. Indeed, we could argue that in this area it's become socially acceptable to indulge in this sort of conversation. They can be classified as statements made to avoid being hurtful or to be socially polite without offending; or to protect someone from the truth. Just think how parents have to conceal information from young children because they're 'not ready' to be told the true facts. Or when those same children in their teenage years conceal details of their 'liaisons' with the opposite sex – because they, the parents, are 'not ready' to be told the truth (there's a switch!). It's a fact that a lot of relationships in personal life are only sustained because of, as Paul Ekman put it, 'the myths they preserve'.

When questioned, by far the top two areas in which people seem to want to become more skilful – in body language terms – relate to:

- **liking** (which we will cover in Lesson 7): *'How do I know if he/ she likes me?'* and *'What makes him/her like me?'*
- **lying:** *'How do I know if a person is telling me the truth?'*

Types of deceit

The spectrum of telling untruths is broad – from not doing homework to 'social lies' in response to the 'How big is my bottom?' and 'Do you think I've got a good singing voice?' variety, to the antics of businessmen negotiating a high-power deal, to the Grand Jury testimony of former President Bill Clinton. But even the best forensic minds have agreed on one thing – there is no 'one thing' that provides the 'Eureka' moment that tells you, 'Aha – that's a lie.'

Thankfully most of the 'lies' that we tell (or are on the receiving end of) are more of the **social** type rather than those that destroy the fabric of society. Telling your dinner party hostess that her home-made crème brûlée was so much better than 'shop-bought' – even though you inadvertently spotted the shop's packaging when you took the bottle of wine to the kitchen – is far better than damaging a relationship. We quite naturally don't want to make people feel bad about themselves on occasions, or to embarrass them.

Just an aside here – sometimes we'll get into the situation where a person is telling us something which is patently a lie but they believe it to be the truth, and they won't budge. Tricky one this, isn't it? What they're telling you, which somebody has *recounted* to them, is clearly a lie to you, but this person believes it to be the truth. For example, the builder says to you that you told one of his workers to fit stainless steel dimmer switches, and the worker maintains that you had said, when asked about the choice of plastic or steel, 'We'll leave it to you.' Somebody's lying or somebody's got false-memory syndrome.

Check the cluster of behaviours in a situation like this because what you perceive to be a lie may actually be a misunderstanding or genuine belief, and so *the lack of negative signals* (which we'll look at shortly) would confirm this. So, if somebody tells you the Earth is flat – and they genuinely believe it – sometimes you may have to go along with it.

We spend most of our lives interacting with people in relationships. If we always said what we thought, society would break down. **For more serious things we rely on the fact that people tell the truth, otherwise that breakdown of trust (which is the most important thing in any relationship) is difficult to repair.**

BODY WISE

If somebody tells you the Earth is flat (and their body language shows they think it's the truth), accept it – *providing they promise not to push you off the edge.*

How difficult?

The consensus seems to be that unlike most other aspects of body language, **lying proves very difficult to read.** This applies to laypeople, such as parents, and people in occupational roles, such as the police, judges and jury, and politicians (even though many are quite adept themselves!). Tests by Paul Ekman in the 70s and 80s showed that when people were asked to detect behaviour that was indicative of lying, the results repeatedly came out at *no better than chance* – in other words 50/50.

As with the reading of all other aspects of body language behaviour, there is no one single gesture that you can look for to confirm your views. It's made worse by the fact that it's an activity – since we're talking about lying, let's be *honest* about it – that we've all been practising since childhood. Even as teenagers, on the cusp of adulthood, we'll be selective with the truth with our beleaguered parents. So – practice *made* perfect.

So, you have to work those observational skills a little harder and sharpen up your listening skills in the area of vocal inflection.

❝ *Sharpen up your listening skills* **❞**

Baseline behaviour

If there's one golden rule that you take from this topic of trying to detect lies from a person's body language and vocal cues, it's an understanding of the importance of *baseline behaviour*. This is important if we're concerned about the reliability of our observations.

Of course it's easier with people you know intimately and people you have frequent interaction with, both personally and professionally. Friends, family and people with whom you have frequent contact will have provided you with a subconscious 'store' in your memory of their baseline behaviour. You're aware of their general demeanour in a 'normal' situation when they're speaking truthfully, and so a departure from the 'norm' may alert you to a discrepancy between what they're saying and what you can see – or hear in terms of 'paralanguage'. But you can also quickly establish a pattern after meeting people for the first time. Again, the watchwords – look and listen. You're observing bodily behaviour (obviously including the face) and vocal cues. After a short time you'll have two sets of behavioural cues to interpret. So much of this is down to your all-important 'intuition'. That subconscious store of knowledge that the brain registers and stores and then provides us with a warning sign if behavioural cues are different – a kind of 'sixth sense' or heightened awareness in reading behaviour.

Observe a person's behaviour 'style' when they are telling the truth about factual statements. We all have our own personal idiosyncrasies. For example, some people may have excessive eye avoidance when speaking, or they may just move about in their seat frequently or fidget a lot. So to conclude at the outset that their actions betray guilt, anxiety or deception may be totally inaccurate. If that's their 'baseline' when responding with truthful answers then these are obviously not good indicators in detecting deliberate deceit.

If you're dealing with somebody you already know then it's a little easier. You can compare a person's baseline behaviour with any deviations (again, that occur in clusters) that are displayed. If it's somebody you don't know too well or are meeting for the first time, then observe their manner of behaving and speaking **when everything is relaxed and easy-going** – in other words, when there is no discomfort on their part.

As you begin to become familiar with a person's mannerisms, look at all the areas we've discussed in the previous Lessons for indications of 'deviant' behaviour. You need to look at facial expressions, eye activity and gaze, hand movements, self-comfort gestures, arm movements, feet and leg activity, and paralanguage.

No one ever said that detecting lies was easy. There are too many variables at play. An individual action or gesture may be exhibited for a reason that has nothing to do with deception – merely discomfort. So – more than ever – we have to look for clusters of behaviour in order for us to make an evaluation.One expression or gesture cannot be used to detect deception all the time. A rub on the side of the nose, for example, after you've asked an interviewee whether they have any other job offers 'on the table' may just be a spontaneous action at that moment – on their part – as opposed to a 'disguised mouth cover' as they blurt out an untruth. We may strike lucky on many occasions but at other times our interpretation will lead to an incorrect conclusion.

❝ *One expression or gesture cannot be used to detect deception* **❞**

Even with people you don't know and are perhaps meeting for the first time, it's possible – after even just a few minutes – to subconsciously register their normal modus operandi in behavioural terms.

Life is like a game of poker

If you're familiar with poker (or you've ever been accused of having a 'poker face') you'll know that the excitement of this ever-popular game has a lot to do with **the reading of your fellow players' body language**. *But of course they're doing it to you too.* So you have to be expert at controlling your non-verbal behaviour so that you're able to hide your feelings (minimising leakage). Then of course there is the bluffing – putting on the body language to indicate you've got a bad hand (when it's good) and when you've got a bad hand, the opposite.

You probably engage in a lot of poker-type activity at work – especially if you work in an office and you attend meetings as part of your job. As we discussed earlier, **it's in the workplace that you see the most 'masking', as we all play our respective 'roles' at the same time as trying not to display any hint of weakness in our façade that may indicate we're 'not up to the job'.** Substitute the poker table for a meeting room table and you see the same 'game playing' at work. In any negotiation the same tactics of hiding true feelings and bluffing come into play.

You may have heard the word 'tell' in poker parlance, which refers to a particular body language gesture transmitted subconsciously (leakage) – the way a person twiddles with their pencil when they're anxious, for example – or deliberately, as part of a bluff. You may have seen films in which one or more players are wearing outsize dark glasses so as not to give away anything through the eyes. **We know that pupils dilate when we're experiencing joy of any kind. Also the blinking rate increases in times of anxiety or anger.** Whether you play poker or not, it's interesting to watch players in action, whether it's in Las Vegas or your living room. As well as observing a great deal of masking of feelings and the challenge of spotting if you've witnessed a 'tell', you'll see a good demonstration of how liars pit their wits against chance.

Back to the real world. We can't know just from *one isolated gesture* if a person is deceiving us (for whatever reason, be it harmless or more serious) – we need to look for a number of supporting clues. To help us we'll take a look at all the research findings, and the most effective clues as to whether a person is being 'economical with the truth' or whether they're telling a downright lie.

BODY WISE

Studies have shown that as many as 90 per cent of lies that are told produce telltale signs through the body or 'paralanguage' (vocal).

The microexpression

The face is obviously the main focus for covering up deception. When there's a clash of feelings or emotions, the turmoil in the brain means that the canvas of our emotions – the face – may display that flashing 'microexpression' that Ekman introduced us to after his extensive studies on the human face. Being involuntary, the idea is that they express the true emotion felt, as opposed to the emotion the person is trying to present outwardly. The limbic area of our brain – which handles emotions – throws out a spontaneous physical reaction if we're feeling any kind of negative emotion – guilt or shame, for example.

This microexpression comes and goes in a fleeting moment – it can last for less than half a second – before the 'mask' takes hold again. This mask could have been one of concern, joy or even a 'poker face', but it's interrupted by the microexpression of a sneer, for example, or malevolent smile or other contradictory expression – very revealing for your best friend, boss, spouse, police or jury in a court.

❝ *This microexpression comes and goes in a fleeting moment* **❞**

These brief facial movements – that leak the truth – are involuntary and only the really observant and perceptive person can pick them up. The small or inconsequential lies that a person tells, 'I only had *one* drink at the station bar', don't usually provoke enough emotion to generate negative 'leakage' (though perhaps another type of leakage in the train's WC!). Lies of greater magnitude usually generate enough physiological activity so that it shows on the exterior.

Telltale smile

There's a general perception that you can tell when somebody is lying about something because they will tend to display a smile in order to mask the truth. The reason being, of course, that it's the polar opposite of a person's countenance that you would expect to see if they were telling an untruth. A smile looks less suspicious and also has the added advantage – as we know – of usually generating positive feelings in the other person. That's what most of us human beings are like – always taking people at face value, aren't we?

All the studies show that in the main *it's the opposite* – these people display *less* smiling. It's not that the liars don't smile: **they just smile less**, believing that because the common perception is that people expect them to smile a lot if they're lying then *they'll thwart them by not doing it*.

BODY WISE

Contrary to what most people believe, research studies show that a person who is telling a lie will display less smiling than somebody who is telling the truth.

When people are smiling they'll adopt a 'false' smile while they're lying (lower half of the face, remember?). Again, look for the classic signs of the non-genuine smile (discussed in Lesson 2). Remember that the fake smile appears *rapidly* and is held much longer than the genuine one, and then rapidly disappears from sight.

The 'felt' or real smile, just to remind you, appears *slowly* and then fades slowly as well. It's tough to produce a real smile when you're actually lying to someone. **Also, remember that the non-genuine smile is asymmetrical** and the corners of the mouth, rather than being upward, are turned down. A lop-sided smile is, for most of us, quite evidently not a true smile of joy. But we don't tend to care much in normal 'polite' interactions because a smile is a smile and we're conditioned to appreciate that. But if we're looking for clues to back up other telltale body signals, the type of smile will help us make our assessment.

Eyes

In counteracting what is normally thought of as deceptive behaviour associated with the eyes – 'shifty-eyed' is a phrase that often comes to mind – a person engaging in deception may do the opposite of what is normally expected. In other words, they may use excessive eye contact. Because gazing is a conscious activity, the eyes can be used by someone telling a lie to try to denote sincerity. Ekman, in his research, found that certain Machiavellian types – such as confidence tricksters and psychopaths – engaged in this behaviour.

This is the complete opposite to most people's perceptions – surveys carried out worldwide, over the decades, show that most people *believe* that the number-one give-away is that *people don't look at you.*

To add further weight to this generalisation I tried at a seminar an experiment with one of the women present, using piles of playing cards, to try to ascertain if I could tell when she was lying in response to my questions to her. Long story cut short

– there was fleeting but intense eye contact that gave her away at one point in the questioning.

On asking the others who were present (some of the sharpest minds around, I have to say) as to how they thought I knew she was lying at that point, the answer was 'the eyes'. Well their response may have been correct, but they said she didn't look at me when she answered. *In fact it was the opposite.* That was the *only time* she made eye contact. At other times she had looked downwards or was giggling with the person who sat to her immediate left. All the other people present remembered her not looking at me at the point when I spotted the right answer. Confirmation that most people associate *looking away* with deception.

66 Most people associate looking away *with deception 99*

There's no doubt that you'll come across people who won't look you in the eye when they're telling an untruth or when they're being asked an awkward question appertaining to dishonesty. They may be looking downwards all the time or their eyes may be darting here and there. Rather like what children tend to do when questioned by a parent or other adult. They too can go in for excessive eye contact as well when they try to convince you of their innocence. Again, you have to look for supporting body or vocal information.

We've spoken earlier about 'baseline' behaviour. *This is one of the terms you should overfamiliarise yourself with.* If you can identify someone's normal way of operating, then it's easier to notice deviations from that. For example, if someone's *normal* way of conversing is with minimal eye contact (due to shyness perhaps) and when questioned or pressed about something they start using unusual and prolonged eye contact, then some alarm bells should be ringing.

Check their blinking rate as well. How does it compare to their normal rate – when you weren't asking them searching questions? *Look for changes in baseline behaviour and note when they occur during a conversation – it will tell you a lot.* But also remember that nervousness and anxiety cause similar behavioural 'tics', which is why it's important to look for other clues as well.

BODY WISE

Rather than looking away, liars may overcompensate and use excessive eye contact.

Eye direction

We'll touch on some research that has been done in connection with the direction of a person's gaze, providing you treat these findings with caution. They can be revealing if tested beforehand by observing a person's baseline behaviour. I can't stress it enough – *observe a person's normal behaviour and commit it to memory.* **Then it's a question of looking at a person's body language deviations in order to be able to read them much better.**

66 The direction of a person's eyes can be quite revealing 99

You may recall we discussed earlier (in Lesson 2) that the direction of a person's eyes at some stages of an interaction can be quite revealing. Briefly, and I stress *use this with caution*, it's been agreed by neuroscientists that each side of the brain is responsible for different functions. The left side is the logical side dealing with rational, analytical and linguistic activities; the right side is the more imaginative and creative side, and the more 'intuitive' side.

Following on from this clear division of the left- and right-brain functions:

- When you're searching for the answer to a question from the 'storehouse' of information that's lodged in your brain (in other words, recall) your left side will be more active.

- If you're lying then there is no recall, but there is a need for imagination in order to create a fictitious answer, so it's right-brain activity.

It has been discovered in research that since we all naturally break eye contact while we're speaking to another person, *the direction of our gaze* can reveal – I must stress, after we've established 'baseline' behaviour – whether or not a person is telling the truth. How?

Each side of the brain controls movements of the *opposite* side of the body (something you may already be aware of). So, just to be clear, the left-hand side of the brain controls the right-hand side and vice versa.

- If a person gazes to the right while they're answering your question, they may be telling the truth (the left brain, as we noted, being responsible for recall).

- If they gaze to the left, they may be fabricating an untruth because it is originating from the imaginative right-side of the brain.

Once you've surreptitiously established a pattern for a person's eye direction, it can be a useful indicator to help you – along with other clues – to establish honesty.

CAUTION

As I stressed earlier, this theory needs to be tested using a previously established baseline. There are some people whose baseline behaviour may be one of gazing in the same direction all the time – purely an individual habit. (Equally, you need to establish whether a left-hander reverses the process or not.)

Blinking and 'eye rub'

A person's blinking may **increase** when they are lying because the increased cognitive activity associated with dealing with stressful questions accelerates everything. It's not always because of lying, it happens when you're under pressure or tired.

It has also been discovered that there's a certain type of person who engages in what is known as an *eye rub*. Looking like the activity that's associated with fiddling with contact lenses (be sure it's not that), **the person will rub their eye(s) while delivering the lie**. It seems to be common in both men and women, with men producing a more rapid action and women confining it to a touch to the side or below the eye. It appears that the greater the perception of the lie, the greater the associated activity with the rub. So it may be accompanied by looking away from you.

What's the psychology behind it? It's accepted that it's a kind of blocking out of the deception, or of the person to whom they're dispensing the untruth. We're under threat and we want to erase the image of the person in front of us.

Face touching

Hand-to-face displacement activities are, for most people, quite prevalent when lying and there is usually an increase in their use. **We all, to varying degrees, touch the face during our interactions with others, but research shows there is much more of this activity if there is deception involved.** Again, tread carefully, there is no one particular gesture. Use it in combination with other clues to recognise a cluster and this will enhance your efforts in recognising a lie.

❝ Use it in combination with other clues ❞

Mouth

For children, just putting a hand in front of the mouth is usually enough for a parent to know that a lie is about to be told. As adults, we're more sophisticated in the magnitude of deceit and the ways in which we try to prevent it leaking out. **However, we may still need the comfort of putting the hand to the face, especially the mouth.**

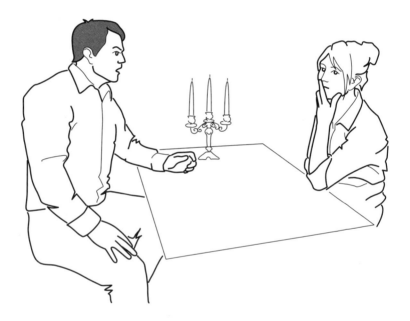

When we're surprised about something notice that we'll instinctively raise a hand to the mouth and cover it. We do this subconsciously to guard against letting out a knee-jerk emotion as the mouth is forced open – we need time to evaluate. **So we're still wired in adulthood to bringing a hand to the mouth in times of stress or surprise.** In a situation involving a lie, the hand may go up to the mouth (commonly known as the *mouth cover*). It's a way of covering up a lie – or something you feel you shouldn't have said (too late!). The brain is rejecting the false message and instructs the hand to do a cover up. The palm of the hand usually covers the whole mouth, with the thumb pointing towards the ear. (It sometimes gives the person an anxious look – if not an outright lie, it usually denotes some kind of anxiety or doubt.)

Alternatively, it can be a hand *supporting the chin* with a finger just touching the mouth – this is similar to the 'shh' that we were used to as children, when you hold a single finger against the lips imploring others to be quiet. Our subconscious is sending a signal to the hand telling us to be quiet.

Another position is *all of the fingers fanning the lips.*

Sometimes it's a *fist that covers the mouth* (See next page) as opposed to the open palm. **This is seen more often when the listener has heard you speak and thinks you are concealing something** – they'll adopt a hand-to-mouth position because they're suppressing a subconscious urge to challenge you.

A word about lips

While we're on the mouth, we should talk about lips. Not the touching of lips, which we've covered, but the activity of the lips independent of that. We spoke about lips in terms of different smiles. You'll remember that there is a complex network of muscles around the mouth area that can all behave independently of each other, allowing a myriad of expressions. As such, they can reveal a lot about our emotions (tension is a big give-away) and also conceal ('stiff upper-lip').

We all, at one time or another, display *tight lips* – **the tension usually means we're trying not to display our guilt about lying, our anger or secret delight about something.** Equally the oft-quoted 'stiff upper-lip', in which it is just the lip muscles above the top lip that tighten, is used to suppress a particular emotion, usually, of course, negative. It can give the person an air of suspicion, since it is indicative of concealment of some sort.

A particularly interesting activity, that quite often occurs during a state of anxiety, or with a cluster of lying behaviours, is *the biting of the lip*. We know from what we've just discussed that in stressful situations we increase hand-to-mouth behaviour **and may also chew on things or bring objects to the mouth.** The lips, of course, are part of that very same mouth, so it's small wonder that we'll use them for comfort. You'll see a bite of the upper lip as the lower teeth come forward and touch it, or a lower lip bite where the upper teeth come down and bite the lower lip. Again, it's usually part of a cluster of anxiety activities. Back to one of our 3 Cs again – context.

Of course it's evident in other situations when you're told to 'bite your lip' – in other words, when we want to avoid coming out and saying something in order to 'keep the peace'. You've probably adopted that philosophy at work many times, as well as with your nearest and dearest.

It can also be seen when someone is trying to convey **empathy** and there are many people in the public eye that use this almost as a trademark display. Bill Clinton – he seems to provide us with a lot of case material – commonly uses the **lower-lip bite** in some of his speeches and interactions with people. Along with his other facial expressions, he's showing that he has empathy and 'feels the pain' of his audience.

Another common activity is caused by *drying of the mouth* due to pressure or anxiety, and so you may see excessive licking of the lips or inward sucking of the lips in a pursed fashion, but this is due to a desire for lubrication.

❝ *This is due to a desire for lubrication* **❞**

Nose

Folklore seems to have endowed the lay person with some concrete proof of lying – firstly (as we've discussed) that the person won't look at you. So – guilty. They touched their nose during a conversation. Definitely guilty. (It's often referred to as the 'Pinocchio' effect.) **If only life were that simple**. By now, it should be patently clear to all of you that if we have other clues, or leakage, we may very well deduce things correctly.

An alternative displacement activity is touching the nose as opposed to the mouth. It provides the self-comfort of *covering the mouth with a disguise*. A hand goes up to the nose and conveniently covers the mouth, which is what you're really after. The mouth cover becomes a by-product of the nose touch.

As an excuse for covering up the mouth, you'll see this gesture repeatedly in everyday life. **However, on occasions you *need* to touch your nose**. Not because of the effects of cold, hay fever or whatever. *That of course is perfectly legitimate.* (However, you should be aware that, as with all body language, it's sending out a signal to the other person who may, in all likelihood, misinterpret it.) You may have to touch it for *physiological* reasons. What do we mean by that?

BODY WISE

Touching the nose relieves internal tension. It's not always associated with deceit because stress with increased blood pressure releases chemicals that cause nasal tissues to swell.

Look at *when* and *how often* it occurs – and more importantly the context.

It's your *autonomic nervous system* at work again. When you're under pressure – you could be telling a lie under traumatic circumstances or just in a pressurised, stressful situation – **there is an increased flow of blood into the tissues of the nose causing them to swell.** The nose grows slightly larger and the inflammation – fortunately not visible to the naked eye – gives rise to that itchy or tingling sensation that makes you touch or rub it, or whatever.

Nose touching gained a lot of publicity in 1998 when President Bill Clinton was televised before the Grand Jury for his involvement with Monica Lewinsky. This piece of 'theatre' has gone down in history for its content and obtuseness. A definition of the word 'sex' was needed, in response to one of the questions. (Previously, during the Paula Jones testimony, Clinton's questioners had to deal with his response: *'It depends what the definition of "is" is.'*)

However, the trial also provided television viewers (as well as the Grand Jury) with a masterclass display of behaviour associated – at the very least – with anxiety. He certainly used the 'mouth guard' a lot, fixed an intense gaze and there was little movement of his hands, which were mouth-directed. As far as the nose was concerned, an analysis showed that when he was speaking about things that were obviously the truth, *he didn't touch his nose at all* – in fact his hands were nowhere near his face during these points. However, while discussing his involvement with Ms Lewinsky *he touched his nose every four minutes or so, for a grand total of 26 times.*

It's difficult to know whether or not the former president, in asking for redefinition of certain terms, had convinced himself that he was therefore not lying to the nation and within those narrow confines, for purposes of that testimony, believed he was telling the truth. If that was the case it certainly wasn't reflected in his performance while fighting for his political life and avoiding impeachment. **The number of repeated displacement gestures, captured by the unforgiving TV cameras, was certainly illuminating**.

Hillary Clinton seems to be afflicted – maybe quite innocently – with a similar propensity for redefining words in the English language. During her Democratic nomination campaign in 2008 she suffered a big setback in terms of integrity when she '*misspoke*' about what happened during a visit to Bosnia.

Hillary had been recounting to the crowds how, during her period as First Lady, she landed in Bosnia in dangerous conditions under heavy sniper fire.

'Then I gnawed through the ropes, **kicked a couple of alligators out cold** and pulled myself to safety.'

Hillary Clinton

'The new *Indiana Jones* movie?'

'No, just Hillary's stump speech.'

Cartoonists had a wonderful time with this (just before the Cannes Film Festival).

Obliging TV stations dug out old footage showing her strolling across the tarmac and smiling as she was serenaded by a little girl.

Misspoke? This was her response, with guarded body language, when asked about a lie she had just told. She hadn't told a lie – she just 'misspoke'.

Hands

We've looked at what the hands do to the face, now let's see what they're typically up to when they're away from this area. **What's noticeable from all the research is that hand actions tend to decrease from a person's normal 'baseline' during deception.** They are *still*, rather than animated, contrary to what most laypeople think.

Normal hand gestures (illustrators) used to accompany speech and reinforce a message are *noticeably absent*. There is a tendency to be on guard and suppress movements – especially of hands – because quite often you're unaware of what they are actually doing when they're accompanying your words.

The hands typically may be hidden. Be aware, as always, that this also occurs when people are just nervous. *There's always been a symbolism between hands and the heart* ('with my hand on my heart') *denoting sincerity,* so the subconscious may instinctively seek out a hiding place for them – under a desk, in pockets or even hidden under the armpits. In other words, creating an arm-cross – that defensive 'closed' body position – which certainly doesn't give a good impression. Alternatively one hand may take a grip of the other and clasp it, pushing it down and holding it in that position.

BODY WISE

Body movements generally tend to be slower if a person is engaging in deception.

Equally, there are times when a person (check 'baseline behaviour') may engage in activities *away* from the body – like excessive fidgeting with the fingers, picking up a pen and twirling it, clicking it up and down (you've done this, I bet), doodling subconsciously on a piece of paper or rotating an empty glass on a table. These 'adaptors' – you'll remember this from an earlier Lesson – may give a clue if they occur as part of a cluster of other clues.

Legs

It may come as a surprise to you but extensive research shows that the area of the body that is best at displaying whether or not a lie is being told is *the lower part of the body,* below the waist.

Solid research (Ekman and Friesen) found that even though the legs and feet are under our conscious control, **because they are the furthest away from the brain** – especially the feet – they are the *least* controlled parts of the body when somebody is trying to deceive.

BODY WISE

When a person is feeling comfortable and feeling confident, the natural inclination is for them to spread themselves out – as best they can. When in a situation of discomfort, the body *closes in on itself.*

Locked feet and ankles

In the lower part of the body, ankles may be crossed around each other in the familiar **locked ankles** position in a stressful situation. The interesting thing is that in this situation the feet do not move much and this may accompany the restricted hand and arm movements that often occur with this cluster – closed body language. You may have noticed this, often with women in dentists' waiting rooms as they nervously flick the pages of two-year-old magazines. Or passengers on an aircraft just after boarding and waiting for take-off. It is extremely common in a job interview situation both while waiting to be called in (not good – you're on show) and during the interview – but it may not be so serious because the *barrier* of a desk may provide camouflage in many instances.

❝ The familiar 'locked ankles' position in a stressful situation ❞

BODY WISE

There's more 'leakage' in the lower half of the anatomy when people are lying.

A variation

Sometimes you'll see a variation of the above position, in which
the person actually **locks the feet around the legs of the chair**
(an unnatural position for men). This is often accompanied by
a tight gripping of the arms of the chair – if they have them –
otherwise it may be accompanied by an arm-cross position or
there may be excessive fidgeting. This is a common seating

posture in situations in which the person is holding something back (or when anxiety is being experienced or they're being defensive for whatever reason). **Their mind and body are in harmony in the sense that the position reflects their negative emotions.** There's a lack of movement because they won't budge from their defensive position.

Movement

Equally, as in the previous example where there is lack of movement, there could be movement of a fidgety nature, which gives a similar message. **Many people sit in a crossed-legs position, for example, with the foot of the crossed leg moving from side to side, up and down or twirling around in a circular fashion.** Some people don't typically exhibit this movement. Watch for a *change* in foot movement from no movement to sudden movement; equally from, say, circular movements to kicking up and down, which usually betrays annoyance or extreme nervousness (without conscious awareness).

This may be accompanied by occasional crossing and uncrossing of the legs. The movements, rather than looking natural, tend to look stilted, as opposed to the more relaxed way in which they happen during normal conversation.

- So look to see if and when this movement with the feet starts. (Not so easy, as we noted earlier, in situations with a barrier in front of you.)

- If you establish that the foot movement occurs *specifically* when you're *discussing a certain point* and then *stops* when you've moved on to something else, that may provide you with certain clues. (Mothers will be very familiar with this *stop–start* motion when questioning their teenage daughters.)

BODY WISE

If you can establish a 'baseline' in foot behaviour then you can identify a change in emotions.

- Sometimes you'll encounter a person whose normal baseline behaviour when seated with crossed legs is one of repeated movement – in other words, when they're *relaxed*. When asked, or they're speaking, about something uncomfortable, the foot movement may actually *stop*.

- So, in this instance, it becomes the *lack of foot movement* that's a clue. *They've gone from comfort to discomfort.* It may not necessarily indicate lying but the point is that it's a reaction to something that has stimulated a negative thought.

At other times you'll observe that the body may slump down and the feet may be tapped, or body posture moved *away* from the source of discomfort with the feet (or one foot sometimes) pointing towards the escape route – the door.

Under cover

In the average workplace, in meetings and during other interactions, desks provide a 'shield' for most people and so it's just the activities of the top half of the body that are observed. **An awful lot of nervousness may be going on below – under cover.**

Think of job interviews you may have been part of in the past. The protective shield of the interviewer's desk may have helped you in the sense that you spent your time projecting a confident image from the waist up. Or maybe you've been to a panel interview where your whole body was on show. How different was that?

❝ The protective shield of the interviewer's desk ❞

In restaurants, what's going on below during that business lunch while you're negotiating, or during that romantic encounter? That confident, no-nonsense, final offer with the seemingly unflinching client or potential boyfriend. *What's their body language betraying under the desk or table?*

What about law enforcement agencies – the police and the like? You've seen it on TV and at the cinema – the person being interrogated with a full body view. Why? *To get additional information from the lower, least controlled part of the body.* The fact is – the underrated legs and feet reveal a lot.

If you're standing when you're interacting with someone then obviously you'll be able to observe whole-body behaviour and also shifts in posture at specific times. You may witness a person tapping their feet, and if they're not accompanying it to the lyrics of 'Mamma Mia' – in other words there is no music in the background – it **suggests some discomfort** and that they would like to flee.

It's usually in a workplace setting – where we have desks to hide behind – and also dining tables in our personal lives that we're able to prevent the lower half of the body talking too loudly.

Vocal

The second aspect of non-verbal behaviour – the non-verbal aspect of speech – provides us with important clues which, along with other body clues, can reveal much. **In fact most deception or withholding of truth is given away by people's speech.** Not just by content but also by the *way* in which things are said – the area that we are interested in, of course, is *paralanguage*.

Speaking slower

The words may come out more slowly than the normal rate because the person is having to rely on memory, rather than the truth. In addition, there's a lot of cognitive activity going on because the brain is being asked to juggle the truth with the lie and also to avoid leakage through the body. If there has been a great deal of rehearsal then, of course, it's possible that there might not be a detectable difference in the rate of speaking.

Silences or pauses

There may be discernible pauses or silences at specific points during an explanation. Deception usually produces more pauses than usual between individual words and sentences. There may also be a trail of abandoned sentences whereby a train of thought is not finished – and then there's a long silence and the person doesn't go back to the original point but starts a new sentence.

Breathing pattern

A person may be trying to adopt a relaxed posture while experiencing discomfort – whether they're standing or sitting – but the telltale sign of a chest rising up and down is usually impossible to conceal, especially if there is shoulder activity. As the breathing becomes more shallow and rapid, this, coupled with voice change when speaking, gives away a lot.

Throat clearing

We do it all the time. Swallowing may become a problem because the stress reaction causes a dry throat. However, an excess of this activity indicates extreme discomfort.

Speech errors or hesitancy

There may be a lot of 'errs' and 'ums' punctuating the conversation, instead of a free flow that is not reliant on fabrication.

Pitch

Like every other pointer we've discussed, you're looking for a *change* in a person's regular vocal characteristics. Pitch can be a very good yardstick to indicate a change in a person's emotions because these tend to cause a rise that is very difficult to hide. So the voice tends to come out higher, and maybe louder, compared to the baseline.

It can't be stressed enough that because there is no one single clue for detecting deception, it's important to take the all-important auditory clues – if you feel they are significant – and combine them with body clues to see if there is congruence (one of our 3 Cs).

❝ There is no one single clue for detecting deception ❞

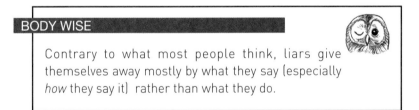

BODY WISE

Contrary to what most people think, liars give themselves away mostly by what they say (especially *how* they say it) rather than what they do.

'Below par' performance

The world had to wait a few months for golfer Tiger Woods to make a public confession and acknowledge his extra-marital affairs. If his apology was to be successful, it had to be believed.

He did this in February 2010 not to a press conference, but to a hand-picked audience of family, friends and business associates at the PGA Tour's headquarters in Florida. It was imperative that his sincerity shone through whatever he was about to disclose. His body language and his words especially the delivery

– were going to be key to the impression he was about to give to the millions watching on TV.

The first thing people notice is the appearance of another person. The golfer looked as though he was there under duress and although normally an elegant dresser, he wore an ill-fitting shirt without a tie and it looked as though he had dressed hastily to be in time for an uncomfortable appointment. The following probably paints the picture:

Tiger Woods makes his entrance

Stage-managed show reveals a petrified superstar in freefall

(James Borg, *Daily Telegraph*, 20 February 2010)

. . . Woods will be returning to therapy, and on this evidence he is certainly not ready for a return to golf. His first actions on making his entrance – concentrating on looking downwards, arranging his notes on the lectern before levelling his gaze and greeting his audience – were typical of a man who is both tense and dejected.

He gulped frequently and his eyes shifted from left to right, both indicators of severe nerves. It was surprising that he did not convey more sincerity with his delivered statement – especially as he appeared more concerned with his words than anything else.

How words are delivered is a big part of body language (paralanguage) and it is easy with excessive concentration on words to lose control of tone. Woods spoke like an old-fashioned 'speak-your-weight' machine, his tone even throughout.

One would have expected him, especially when speaking about his family, to have dispensed with his notes for a few minutes . . . he continued the pattern of speaking for 10 seconds, referring to his notes, and then speaking for 10 seconds again. It was mechanical and raised the element of doubt over his sincerity. . . . A wooden delivery . . . could have conveyed a lot more had he remembered the time-honoured maxim in sports and showbusiness – image is everything.

Unfortunately, during the 13 minutes the only time his voice expressed any emotion was when he lambasted the press for their excessive interest in his family.

BODYtalk

Q So there's no foolproof method for detecting when someone is lying to us then?

No – just like in other situations we have to look for clusters of behaviour to help us.

Q Is it because the symptoms of lying are similar to a person just feeling nervous?

Yes, that's partly it. But you can at least know when a person should not be feeling nervous in a particular situation; therefore a number of signals associated with deception can be looked for.

Q So you were saying earlier that a lot of lies are harmless in the sense that they are social lies – are you saying then that these are quite legitimate?

Well – in an ideal world when Aunt Florence gives you yet another six-pack of socks for Christmas, you could – if you were stone-hearted – give an honest response like: 'How many feet do you think I have?' Alternatively, you could have some empathy and think that the poor lady walked to the store, queued up to pay and wrapped them up – just for little ol' you. So, as in many other instances, is it worth upsetting somebody who doesn't have a clue?! (I never said that. You did.)

Q So it's the more harmful, or shall we say more revealing, lies that we're all obsessed with. Is that right?

Yes. In body language terms, when people are asked:

- **why they want to have a better understanding of non-verbal behaviour**

- **what they want the magic pill to enable them to do**

the two reasons that come out time and time again are – I want to know how to tell:

- whether a person likes me (or dislikes me)
- whether a person is telling the truth (or is lying).

Q Is there a 'must' that you can advise us on that will help in detecting when somebody may be telling a blatant lie or withholding the truth?

Try to recognise what we call 'baseline behaviour'. How do they normally conduct their interactions? Are they expressive, or the opposite? What's the voice like when they are normal? What's their typical eye contact like? Sometimes, of course, you haven't met somebody for long and so you haven't had enough time with them. But even in short interludes you can gain a lot of information as to a person's baseline by asking factual questions. Observing their mannerisms and voice when 'normal' helps you to see how it deviates – if at all.

Q I'm a bit concerned about my feet and legs now. Should I try to hide them?

Well, not if you've got nothing to hide – I don't mean your feet of course, I mean nothing that would induce your nervous system to leak out some unwanted body movement. On the other hand, even if you're not lying about something but you're nervous about being in a particular situation, your feet and legs may do the same thing – as if you were under stress because of lying.

Q Oh, that's okay then. The people present would know it wasn't because my telling a lie, but because I was just nervous.

They wouldn't necessarily. They're witnessing something that could indicate either. They should look for 'clusters', of course, to support it. But if they haven't read this book, well . . .

I see your point.

Q The person before was worried about their feet and legs. I'm worried about my voice. If I'm under stress it changes. I don't want people to think it's because I'm telling whoppers.

Take a deep breath and slow your speech down. The pitch, which has a tendency to rise, will then gradually start to drop towards normal level.

Coffee break . . .

 There is no one single gesture to identify if a person is engaged in deceit. If possible try to observe baseline behaviour and then look for changes.

 Research shows that as many as 90 per cent of lies produce telltale behaviour through body and vocal cues.

 It's generally thought that a lying person reduces eye contact, but quite often a person may overcompensate with excessive eye contact.

A person may engage in a lot of hand-to-face gestures when lying – but caution is needed because the same activity is displayed during nervousness.

The lips are often a good indicator – a person may adopt a tight-lip pose as the subconscious tries to withhold the truth; also there may be biting of the lips activity.

 There may be excessive touching of the nose – there is a flow of blood to the nasal tissues under stress causing the nose to swell (difficult to detect with the naked eye).

 Hand actions tend to deviate from the 'baseline behaviour' during deception – there may be some external activity away from the body like fiddling with objects or tapping fingers.

The legs and feet are quite revealing as they are the least controlled part of the body, being furthest away from the brain.

The vocal aspect of speech, the paralanguage, undergoes many changes during deception.

 Contrary to what is generally thought, most liars give themselves away by what they say, rather than what they do (the words and how they say them).

Lesson 6

'You're not paranoid – you're the opposite of paranoid.
You suffer from the insane delusion that people
actually like you.'

Woody Allen

Leakage

By now, I feel you're super-aware that body language can reveal an awful lot about messages that people are sending. 'Leakage' is the term for body language that gives away what you're really thinking, as I'm sure you know by now.

The mind and body link

The mind and body are inextricably linked. *So every thought produces a body reaction.* Any message can be better interpreted if non-verbals are read successfully. Equally your own body language, as we've repeatedly hammered home, is crucial to how effectively your message is received by others.

It's not the displaying of positive thoughts that we tend to try to conceal – it's usually a state of discomfort or anxiety. No matter how hard we try, in certain instances our body language will let us down, and the 'leakage' means that a message can lack 'congruence' between **what is being said** and **what the body shows**.

When we talk about anxiety there are, as far as psychologists are concerned, two main types exhibited by most of us:

- **trait anxiety**
- **state anxiety**.

No surprises here with the first one. Within our complex personalities, some people could be classified as perhaps being 'anxious'. **It's a trait within somebody's personality** – in other words; it's non-specific; a general characteristic.

State anxiety refers to a person's response to stimuli. So, in a **particular situation** a person will experience anxiety, but at other times in the 'switch off' position there is no such perceived threat, so everything is in equilibrium. Thus we'll experience anxiety or nervousness if we're asked to give a speech or make an appearance on television, for example. It's perfectly normal. Even experienced screen actors can experience extreme nervousness when appearing in front of a live audience, on a chat show for example.

❝ Even experienced screen actors can experience extreme nervousness ❞

Oh... Oh... Seven

Daniel Craig

I'm reminded of the appearance of actor Daniel Craig, after the film release of his portrayal of 007 secret agent James Bond in *Casino Royale*. Appearing on the *Parkinson* chat show, a week after the release of the film, it was interesting to watch the displacement and comfort gestures that were in action.

Unfortunately, it's all magnified on screen and you have to have a degree of sympathy for these people when they're put in the limelight. I made an analysis of this interview and it should bring home to you how subconscious this 'leakage' is – and the impression it gives. I've chronologically itemised the 'displacement' and 'comfort' activities – as they occurred – to highlight the pattern of behaviour exhibited, and how they formed the all-important cluster indicating nervousness.

Sitting down opposite Michael Parkinson, first of all Craig *gripped the arm of the chair* with his right hand. After being welcomed he began speaking (with crossed legs), then he did the following as the interview progressed:

- stroked ear
- touched bottom of nose
- made 'intention movement' associated with leaving (i.e. gripping both arms of chair)
- touched bottom of nose
- scratched side of neck
- moved tongue around lips
- three simultaneous eye cut-offs (when questioned about his mother's attitude to his acting)
- touched eyebrow
- touched bottom of nose
- moved around in chair
- smoothed hair at back of head
- touched bottom of nose
- rubbed nose
- moved tongue around lips
- smoothed hair at back of head then, with same hand movement, did a nose rub on the way down, before gripping arm of chair again
- moved around in chair
- rubbed nose
- rubbed nose
- made 'intention movement' associated with leaving
- touched bottom of nose
- scratched left cheek
- touched bottom of nose.

Michael Parkinson then uttered the word **'finally'** (for the last question) and Daniel Craig **oriented his body towards** Parkinson – to the right – for the first time; **looking more**

relaxed. There were no more gestures in the final two minutes. Only the 'intention movement' to leave the chair at the end of the interview (only this time it was for real!).

❝ There were no more gestures in the final two minutes ❞

Just two interesting points to add to that:

- During the whole interview, Craig's eye pupils were constricted. Due to his pale blue eyes it showed up conspicuously. As we discussed earlier, when you're experiencing something pleasurable the pupils dilate and become much bigger. The opposite happens during times of discomfort or nervousness.

- Also, it was noticeable that when Craig's right hand left the right arm of the chair when he occasionally gestured during his speech, when it returned to that position it was always another grip. This remained throughout the whole interview.

All this was noted, of course, only when the camera was on him during the 14 minutes of the interview.

Doesn't it just go to show that no matter how experienced or famous we are, the true feelings that we're experiencing will always leak through?

BODY WISE

Any time we're feeling anxiety of some sort, displacement activities and comfort gestures come into play. *The body's physiology is unforgiving.*

It's all in the head

What about sporting personalities? How does their body language aid them in their pursuit of success? Let's take a look at that most 'psychological' of sports – tennis. A game that is always referred to as being one that is played in 'the player's head'.

Confident body language in this sport sends a message to the opponent and at the same time influences the player's *own* mindset. So therefore it impacts the player as well as the performance of their opponent. Negative 'leakage' that's displayed – on the other hand – can benefit the opponent and further demoralise the player.

On court, the opponent picks up conscious or subconscious signals from every action the other player displays and it gives an overall impression of whether the person across the net is feeling confident and in control or experiencing frustration (maybe due to their serve that has 'deserted' them), nervousness or getting increasingly tired. How many times have you passed a TV screen on which a tennis match is being relayed and even with the sound muted and no visible score line, you know which player is 'on top'? You can tell just by looking at the demeanour of the two players.

Andy Murray shows his frustration

British number one Andy Murray is a talented player but on occasions – as he all too readily acknowledges – his head is in the 'wrong place' and his body responds with negative body language that benefits his opponent and further demoralises him during the match.

Critics respond that his surly on-court manner and frequent outbursts to himself and his 'entourage' in the players' box prevent him from achieving his potential.

Talking to BBC Sport in April 2011 he said:

> In Miami I was getting more angry . . . I was just kind of lost. It seemed like I was trying to do so many things – without coming to the net . . . You really need to go into a match with a clear mindset. It probably showed in my body language and my mental state.

You've seen players with the following negative type of body language on court, I'm sure:

- slumped shoulders
- head down as they walk back to the baseline or umpire's chair
- shaking of the head after a point
- muttering or shouting to the 'self' or to others
- walking slowly
- breathing heavily.

Rafael Nadal comes back from a losing position

Contrast that with the player who's feeling good and wants to intimidate their tired and dejected opponent. You've seen it – fist pumping, head up, quick and rapid movements, jogging to the chair at the end of a game or set, juggling their feet while seated (can't wait to get back on court). All the time they're aware that this upbeat body language has a great impact on their own game as well as the performance of the opponent on the other side of the net. It's been said that the great champions have emotional control, which separates them from the rest. The self-sabotage body language (smashing rackets and balls, self-abuse) is substituted for emotion that is directed towards their shot-making.

Multiple Grand Slam champion, Rafael Nadal, is renowned for not wanting to lose a single point, even when he's in a commanding winning position. As well as being known for chasing every single shot, he confines his negative body language after a bad shot to the natural show of disappointment – a wince or disappointed expression. Even if he's in a losing position he channels his energy into his shots (rather than defeatist body language) and a winning backhand or forehand is greeted by him with fist-pumping and jumping and encouraging looks to his team of coaches in the seats. He, like many of the other top players, realises that adverse behaviour affects the mind – and therefore his own performance – as well as sending out a message to the opponent that he is frustrated and not in control.

Changes in mind and body

In order to understand negative leakage better, I feel it would be helpful for you if we take a look at negative thoughts or anxiety and **the subtle changes that take place in your mind and body**. These affect:

1 your thinking
2 your emotions
3 how your body feels physically
4 your behaviour.

As you think, so you will feel

Because stress and anxiety originate from a cycle of activity (1 to 4 above) then, of course, if we could reshape our thinking this cycle of negative behaviour – **which reflects itself in our body language** – could be halted in its tracks. But quite often it's too late – you're already at 3. You're in the room being interviewed for the job (your thoughts are screaming, 'There's too much competition, I won't get through this. I did so want this job . . .') and the physical symptoms of your body – pounding heart, muscle tension – are producing this anxiety 'leakage', which is at odds with the calm, coping image you wanted to portray.

In another scenario you're aware of your '**freeze–fight–flight**' symptoms while you're waiting in reception because your body tells you so. This hauls you back to what is causing this – your emotions – and gives you a chance to change your thinking before you go into the room. More easily said than done, you may say. **But since your thinking (good or bad) creates your emotions – which in turn causes your body to react physically – which then manifests itself in your outward behaviour – then changing your thoughts will change your body language.**

BODY WISE

One of my favourite sayings: '*It's only a thought. It can't hurt you*'.

It's not what happens to you . . .

If you recognise the symptoms of anxiety or stress you have a chance of taking control. We spoke earlier about the popularity of '**emotional intelligence**' in which managing one's emotions plays a big part, and of course the self-awareness to recognise these in yourself. So if your success in interpreting body language in others is to do with reading thoughts (effectively mind-reading), it is also concerned with you identifying and being aware of your *own* thoughts, which gives rise to various emotions and therefore your own body language (good or bad).

❝ *Being aware of your* own *thoughts* **❞**

Since it's not what happens to us in life that causes us anxiety, but rather *the view that we take* of the events, we need to be able to control and change these perceptions. We know that negative thoughts fuel anxiety and anger. If you don't think the thought, you don't feel the emotion.

> **The highest possible stage in moral culture is when we recognise that we ought to control our thoughts.**
>
> **Charles Darwin**

Of course, you've often heard of the benefits of the 'power of positive thinking', but in the battle of the gladiators between negative on one side and positive on the other, who finds it easier to win the contest? **Usually it's negative thinking because as the whole stress cycle begins, the depressant activity of the stress chemicals causes more and more upsetting thoughts to crowd the brain, which just makes you feel worse.**

CAUTION

Negative thoughts are self-generating and so they just *multiply*, whereas positive thoughts, unfortunately, do *not* operate in the same way because they tend to be much *less* self-generating.

Try to train yourself to replace any worrying thoughts that are causing you anxiety by more positive ones ('I won't dry up and forget most of the speech', 'I'm just as qualified as the others who have applied for this job'.) The brain can concentrate on only *one* thing at a time, so if you focus your mind on positive thoughts (effectively *supplanting* the old thoughts) you'll feel in a much better mood. This is, to a large extent, the result of 'the power of positive thinking' on the body's *biochemistry* as it releases the good chemicals while at the same time diluting those that were released during your anxious state.

The biochemistry – what's happening?

Just a quick look at biochemistry – the chemical reactions in your brain and body:

- Your thought, which starts in the cortex area of the brain, activates a number of other nerve cells on its route down the neural pathway to the mid-brain area, the limbic system – the home of emotions.

- These then send a message to the adrenal gland near your kidneys, which accelerates the action of many of the other organs that then release a number of chemicals into the bloodstream.

- These chemicals flood the body system en route to the pituitary gland, located in the brain just under the hypothalamus, which in turn releases more of these stress chemicals from the adrenal gland.

- Names of these stress chemicals, I hear you ask? (I hope you're not asking just so that you can justify any abnormal behaviour – such as anger – towards friends, family, work colleagues by tossing out the names of a few stress hormones in defence!) Okay – adrenaline, norepinephrine and a variety of corticosteroids. These are a mixture of neurotransmitters (brain chemicals) and hormones.

> All of this activity plus chemical release activates the body to go faster, especially during fear or anger – but even with mild anxiety the same process is at work.

❝ Even with mild anxiety the same process is at work ❞

No wonder the body gives so much away when we're experiencing nervousness and anxiety.

Let's take a short journey to see what exactly is happening to cause these *outside signals* that give away our discomfort:

- The stomach doesn't handle stress chemicals well. It produces hydrochloric acid so that it can digest your food, but there's no food in this instance. The stress makes acid flow giving you that upset, churning feeling.

- Muscles in your body constrict and under particularly trying situations muscle coordination goes completely and you may shake and tremble.

- Your lungs may become constricted and breathing may become irregular.

- Your sweat glands also become stimulated trying to maintain your normal blood temperature despite the increased blood circulation that's going on. You may get clammy hands or break out in a sweat on the forehead and under the arms.

- Saliva flow decreases and your throat and mouth become dry.

- Stress boosts the blood flow to your skin and this may manifest itself as flushing or a red rash (most obvious during embarrassment).

Effects on the brain

These stress chemicals inflict quite a bit of damage on your normal brain function. **This may help to explain why you sometimes get flustered or impatient with yourself when you're with others** (displaying just the kind of body language and vocal irregularities *you normally strive to avoid*) and you can't remember something or think straight:

- These chemicals actually impede and obstruct normal neural transmission – in other words, the passage of information as it passes through the brain.

- The result is that you don't think clearly and you find it difficult to *remember* – ever come out of an important meeting or interview and found yourself remembering what you were trying to remember on the stairs going down? The brain was unable to process the messages it was receiving at the time and so this affected your powers of recall.

- Both thinking and memory are, to some degree, chemical processes. The nerve cells release chemicals called neurotransmitters, which interfere with transmission when they come into combat with the stress chemicals, with the result that both thinking and memory are affected.

- Also, the constricting of the lungs, mentioned earlier, means that you can't take in enough oxygen, which is so essential to the functions of thinking and memory – any reduction or restriction in the amount of oxygen to the brain results in a failure to function efficiently.

Forgive yourself?

So I hope you all now have a better idea (if you weren't already aware) of **how the lack of oxygen to the brain and neurotransmitter blocking causes certain malfunctions in your behaviour when you're operating under anxious and stressful situations in your personal and professional life.** You find yourself not being able to make decisions, think clearly, concentrate fully or remember things. So if you've wondered why your carefully planned presentation to a group of people at a business meeting, wedding or wherever – honed carefully for optimum vocal and body language effectiveness – results in you standing there with a blank expression, *try not to blame yourself too much.* It's a result of those ruthless stress chemicals knocking out your neurotransmitters, which in turn has a knock-on effect on your memory. **For example, 'forgetting my lines' comes in the top five of greatest fears, whether it be for acting, a work presentation or a wedding speech.**

There will always be occasions when disaster befalls us. What about the description of this unfortunate speaker:

> **His face was blank**
> **the audience was mute,**
> **he'd left his speech**
> **in his other suit.**

Back to that rich source of leakage information, apart from showbusiness – the world of politics.

A researcher analysing a speech during a Labour Party Conference noted that while Tony Blair was speaking to the delegates, the then chancellor Gordon Brown, sitting next to him, engaged in the following (we've got this far now, so let's call it 'displacement activity' since you'll be using this term – as well as being aware of it – when you're let loose): **he adjusted his clothes 25 times; bit his lip 12 times; hand-to-face gestured 35 times; fiddled with his cuffs 29 times; crossed and uncrossed his arms 36 times; looked away 155 times.**

Are you thinking what I'm thinking (test your mind-reading)? The words – records . . . Guinness . . . world . . . book – come to mind.

I just get the feeling that there was some negative leakage coming out there – don't you? And that's without seeing the lower half of the body, the all-revealing legs and feet – what might have been going on there? He didn't want to be there, did he? What do you think?

❝ *There was some negative leakage coming out* **❞**

Summary of 'leakage' and avoiding ID 10T errors

I'd like to highlight – as an *aide-memoire* – some of the gestures associated with discomfort. In other words, negative leakage signals.

Always remember that each gesture listed, *on its own*, doesn't tell the story. A number of these gestures together – **let's say three or so** – will give you a good 'reading'. **Otherwise you'll be committing too many ID 10T errors.**

I'm not going to repeat again what they can signify. I'm confident that you're able to identify them easily by now:

- body closed in on itself
- body or shoulders oriented away from the other person
- arms crossed with clenched fists
- arms crossed and gripping the upper arms
- wringing the hands
- raised shoulders (towards the ears)
- legs and arms crossed
- legs or feet wrapped around the chair
- one foot locked around the back of the knee (while standing)
- ankles crossed while seated
- repeated crossing and uncrossing of the legs
- shuffling or excessive movement of the feet
- foot, or both feet, pointing away from the other person
- repeated swinging of the top leg
- biting of the lip
- tight lips
- touching the nose (or rubbing it)
- touching or tugging the earlobe
- touching the face frequently
- mouth cover with the hand(s)
- chin supported by a hand
- licking of the lips
- frequent gulping
- fiddling with objects
- fiddling with the hair
- holding the chin – or rubbing
- excessive blinking of the eyes
- eyes closing while speaking (eye 'cut-offs')
- eyes moving from side to side
- constriction of the eye pupils
- downcast eyes.

BODYtalk

Q I think with all these 'leakage' signals that I now know about, there's no excuse for not being aware of my own – and also picking them up in other people's behaviour.

Absolutely right. I think you've noticed that with a little bit of empathy – so that you really see when you look and also when you listen 'between the lines' – you'll just become so much more effective. It's not even that difficult, if you have the will.

Q You know what I think the problem was – and I probably speak for a lot of other people here and also, no doubt, for a lot of people out there – we don't *know* that we *don't know*.

Glad you've picked that up.

Q So it's all about image really, as well, isn't it? Things have to be congruent or the action doesn't match the image that we want to project. Our body language seems to be really important in creating our image. Any good tips for maintaining the right image?

That's a tough one – you've put me on the spot. Okay. 'Never try to make a dramatic exit in flip flops!'

Q I was trying to be serious.

Serious? Any more questions?

Q Yes, one last one. This ID 10T error. I don't want to sound stupid or anything. Some people have been giggling – they seem to know what it is. Can you enlighten me?

Sure. Just unravel it.

Q Unravel?

Yes. I – D – I – O – T. It means an IDIOT error!

Q Charming!

No offence to you all. Just be wary. Last question, anybody?

Q Yes, here. You said right at the beginning of the first Lesson that body language was not an *exact science*. Can we rely on something that's not an exact science?

A question for you: Who said *science* is an 'exact science'?

Hey, that's just so profound. Wow. Need to get my breath back. I'll be able to use that for years to come. I'll just have to lie down in a darkened room for a bit.

Lesson

7

'You can tell a lot about a fella by the way he eats his jelly beans.'

Ronald Reagan

Likeability

Studies of a scientific nature, as well as ones carried out by the plethora of men's and women's lifestyle magazines, show that there are a number of common denominators that denote likeability. These cover the whole spectrum of interactions from dealing with strangers and acquaintances, to making friends with work colleagues, customers and clients, to 'affairs of the heart' (flirting and dating situations).

We've all come across those people who seem to be able to 'click' with others the first time they meet. They manage to forge relationships through an 'effortless' way of interacting successfully with people. They come across as likeable and, from then on, the 'halo effect' follows them around.

Your likeability determines your success in both your professional and personal life. It's been proved time and time again that it forms the basis of enduring and better-quality relationships in all spheres. **Better love life and friendships** (these two you would expect), **better jobs and promotions**, and better service from people you are forced to deal with such as doctors, tradespeople, waiters – just about everyone. It's been said that without relationships you have nothing in life, society or business.

How does this come about? When you make other people feel good and they get a positive experience when they're with you, they tend to like you and so their 'psychological experience' is a good one. To put it plainly – people have a better time when they're dealing with a pleasant and engaging person! That's the easy bit – the down-to-earth statement. How do you become that person? Well, you've already started on the path and if you practise your empathy skills it will work wonders in your day-to-day life and at work.

If it's a work context, your colleagues will have better relations with you and, equally, customers or clients may want to choose you over other people. Fact is – that's the way it is. In the business world, people tend to do business with people they like. There are still some people who think that all you need to do is turn up, work hard, be loyal and that's it. Well it's not, unfortunately. Both within organisations and externally when dealing with clients or customers, all the findings show that the basics are not enough these days.

❝ People tend to do business with people they like ❞

For example, research from a cross-section of organisations revealed these top three reasons (in decreasing order of importance) for selecting one company over another:

1   I like them
2   knowledge of business
3   responsiveness.

So all the findings seem to corroborate the fact that life appears to be a popularity contest, no matter what you might hear to the contrary.

If people feel good about themselves when they're around you and the psychological experience is a positive one, you're ahead. So I'd like you to ask yourself this: **'Do I light up a room when I walk in? Or is it when I walk *out*?!'**

Okay . . . what about people you come into contact with? How many of them have that noticeable 'presence' when they're around you? Using what you now know, is a big part of that down to their positive body language? **Chances are it is.** If people have a negative experience when they're around you, they avoid choosing you, staying with you, buying from you, listening to you, helping you. It's how it is and it's always been that way.

First impressions last

We touched on the importance of first impressions. The thing about them is that they tend to have permanence about them – hence the 'you never get a second chance to make a first impression' dictum that we often hear.

Meeting people for the first time means that inevitably the opening encounters are comprised of the proverbial 'small talk'. There's nothing wrong with that. It's how it has always been. (As I always say – small talk leads to 'big' talk.) It means that there is a certain amount of 'gut' evaluation in the first stages of an encounter as we weigh up the other person – and the way we do it is through their body language and non-verbal speech patterns. It's not really the content of the 'small talk' that's so important to whether or not you ultimately feel – 'I like this person'. It's more to do with our two-part 'decoder', remember? **Looking** and **listening**.

When you're in such a situation next time, make a point of asking yourself these questions as you talk to somebody after first meeting them.

- What can you see in the facial expressions? Is there much smiling or is it a barely suppressed negative expression? Are you spotting any occasional 'microexpressions', which we discussed earlier, that disclose true feelings? Is the eye contact steady – with the normal 'dance' of movement – and focused on you? Or does the person engage in abnormal eye contact, perhaps looking away most of the time, or looking around scanning the room for other people?

- What about the gestures? We've covered in great detail the principles of the open and closed body states. Is the body oriented towards or away from you? Are they leaning forward as you talk? Are their head nods in the right places? Do they touch you – on the arm, the shoulder or the hand? Is it done in the normal polite and endearing way to emphasise a point or show empathy?

- And the voice – meaning the 'paralanguage' – does the tone match the body language? When the person is smiling do the vocal cues back up the feeling exhibited on the face?
- Then, of course, the words themselves. Maybe the tone is appropriate for the words, but does the body language show a conflict with the words being said, so that there's a contradiction? Has that now made you distrust the person?

All these things count in terms of initial 'likeability'. It's accepted – from all the research that has been done on the topic, even before we engage in any talk at all – that it is appearance that forms the first impression (reinforced by how we then interact).

More research has been done on the subject of charm, charisma, liking – call it what you will – in recent decades and we now know how to improve the skills associated with these qualities. **The secret to likeability is how we make others feel**.

So what are we saying? That it's nothing to do with you? Of course it is – you have to have the self-awareness and you have to do the work. But the end result is that you make others feel good and therefore they like you. Make sense?

What's the other great secret? The secret of how to attain this likeability quality is mainly *body language*.

Appearance

Let's take a look at the common denominators that come out in the research studies on the subject of liking and attraction. Needless to say, **appearance** comes out top of the list. Within that definition, as you would expect, it includes clothing and grooming.

> **I look alright today – I've been dressed for the telly. Usually I look as though I've scrambled up an embankment after a train derailment.**
>
> Victoria Wood

Your clothing is very relevant to body language because it's saying something about you to the other person – whether you like it or not. Haven't you made instant judgements about other individuals from the way that they were dressed? Were you always right? Or did you not stick around long enough to listen to the person speak much (as is quite often the case!) and be proven wrong?

We know that around 90 per cent or so of the 'first impression' evaluation comes from the visual and vocal elements of body language. So your appearance is obviously part of your non-verbal language. Your 'audience' of one or more is making a judgement about you and forming an opinion in just a minute or less – before you've even had a chance to say one word, in many cases. This is quite often – even though so obvious – forgotten when people are making a conscious effort to check that their own body language is sending out the right signals. Your clothing is of great subliminal importance to the onlooker because the first contact between people is always eye to body.

We use our clothing to present an appearance to the other person that reflects our inner self – at the time. Also we use our outward appearance to 'blend in' with the appropriate situation or to convey a certain image.

Look at politicians when they're visiting workplaces or on trips to other countries – you'll see them in a suit and tie in one setting, no jacket in another, open-necked shirt (minus tie) for another setting, rolled-up sleeves when visiting troops abroad. They know that the impression they make will almost certainly start with their appearance and will determine their likeability and whether or not they will win people over. Either they want to identify with their voting public – or they want to convey a certain image of authority and will revert to the more formal mode of a suit and tie. (Apologies to female politicians – the same rules seem to apply, although it seems that women have more scope for dressing 'smart casual)'.

The instant evaluation of posture and body movement is important along with general appearance. The astute person notices if a person moves with confidence in an upright fashion, and also how the head is held – is it held upright or a bit too far back (a suggestion of arrogance or haughtiness), or is it held down with slumped shoulders? Do they stand in the same way that they walk? That gives a clue to a person's personality and determines how people approach them. Unfortunately the hesitant, shy and maybe introspective person exhibits body language that makes them seem as though they're not interested and are unavailable – the lack of eye contact, the body position that is oriented away from people and slumped shoulders. Their natural hesitance makes them reluctant to transmit any signals of interest because they don't feel they will get any response and so their body language – to other people – is transmitting a lack of interest. Who breaks the cycle?

People say that when meeting somebody they initially notice the person's clothes and then assess subconsciously their physical attractiveness. **So much has been written and researched on the psychology of physical attraction, but let's just go over the obvious.**

We know that first impressions are formed instantly and studies do show that physical attractiveness and appearance undoubtedly, initially at least, have a predictable effect on the judgements that people make about others. In fact, as we know only too well, some organisations may employ people purely for that reason.

66 *First impressions are formed instantly* 99

In day-to-day life, beauty may influence a person at first impression stage, compared to when meeting the more average-looking person. This will therefore influence the person's judgements during an interaction. **But in the medium to long term, people generally base their liking of a person on their other traits and will discount their physical 'ideal' if the body language appeals.** As we're always hearing: 'Beauty is only skin deep' and 'Beauty is in the eye of the beholder'. You might like to know American playwright Jean Kerr's 'take' on this:

I'm tired of all this nonsense about beauty being only skin-deep. That's deep enough. What do you want – an adorable pancreas?

(*The Snake Has All the Lines*, 1960)

Smile

Is it any wonder that this depiction of friendliness which evokes a reciprocal response in other people ranks so highly in the likeability stakes. When you smile you make other people feel good. It doesn't have to be full-wattage all the time. Just sincere. As George Eliot once said '**Smile and make a friend; frown and make wrinkles.**'

Even when things are not going too well and you force a smile that is obviously half-hearted, the empathetic person picks this up and is able to delve into reasons for your mixed feelings. No contest – the smile, at the right time, in the right place, works wonders. Angelina Jolie and Brad Pitt ('Brangelina' as Hollywood insiders have dubbed them) are media 'savvy' and know how to appeal to fans.

Expression in the eyes

This comes out very high in the likeability ratings. We have discussed eyes at length, so it comes as no surprise. Eye contact shows that you're interested in someone, both in general situations and in romantic ones (in which the gaze is held a little longer). The expression in your eyes also conveys a lot and people can detect warmth, sympathy and concern. When you give people attention it increases their feeling of self-worth. You feel the same when you're on the receiving end, no doubt.

BODY WISE

If you make others feel good about themselves,
through eye contact and paying attention, you're perceived
as attractive.

Voice

We looked at the non-verbal aspect of speech in Lesson 3 and saw what an important part it plays in how people respond to us. Generally, if the person's pitch, tempo and loudness is 'in sync' with a person's liking, there's no discomfort – so there's a deepening of rapport.

If you're comfortable doing this, take a tip from the experts. At the first impression stage – regardless of the rhythm and pace you normally adopt for your speech – see if you can match the vocal cues of the person you're talking to. In other words, see if you can naturally follow their speed, rhythm, volume and tone. This has been shown to build instant rapport between individuals. After you've gone beyond this stage, then you can gradually revert to your normal style, but if there are elements of your vocal cues that are distinctly different from the other person with whom you would like an ongoing relationship (be it romantic or professional) – for example, the speed – then try changing it, gradually.

Remember that your voice also reflects your image. Because it's the visual appearance we take notice of first, we then go for the vocal elements of non-verbal language and expect to see 'congruence' from what we hear. You would expect a quiet voice coming from a withdrawn and shy individual. The same voice from Clint Eastwood would not be consistent.

We look for congruence between what we see and what we hear and in the likeability stakes there has to be a match. Marilyn Monroe had a voice that reflected her image and personality perfectly. Jack Paar, her co-star in the film *Love Nest*, remarked: 'Marilyn spoke in a breathless way which denoted either passion or asthma.'

Listening

As with good eye contact, this is another winner when it comes to being perceived as attractive and so is very high up in the likeability stakes. How do you feel when you're talking to somebody and they're looking over your shoulder? When their eyes are facing you but 'there's nobody at home' (they're running their own tapes in their head; self-talk)? There's no 'whole body' listening. Or if they take the floor all the time, and then when it's time for them to hand over to you – and become the listener – they say: *'Enough about me. What do you think of me?'*

Use your whole body to listen and see what effect it has on the other person. Again, a generalisation – it's considered that women are instinctively better than men at showing that they're paying attention and are empathetic. They'll widen their eyes, lean forward, smile a lot and use more head nods. This is a potent mixture for making other people feel good about you.

All the research shows that women are bowled over if they come across a *man* who listens – meaning true listening (as we discussed in Lesson 3). **There's an old joke about a chap who complains that his wife always says that he doesn't listen to her – at least that's what he thinks she said.**

❝*Women are bowled over if they come across a* man *who listens*❞

221

When women talk to other women they tend to be more animated (not always – there are exceptions, of course) and show that they're actually listening. **They're listening with their body.** They nod at appropriate times and may make facial expressions to convey empathy. **Men tend to have a problem with focus**, so the theme of a conversation gets lost and the responses are off-target. In addition, there don't tend to be so many *visual* clues to show that they are actually listening. Women tend to be good listeners so they find, with equal-sex conversations, there's not usually a problem.

When somebody comes across a person who truly listens, they want to gravitate towards them. It promotes empathy and this is a big factor in your likeability, as we've stated before. It helps you to get in tune with another person's feelings so that you can read their thoughts.

Posture and gestures

We've spent a long time discussing the subliminal ways in which we evaluate whether or not a person's demeanour is pleasing to us. The words mean nothing if there's something not right about the way a person moves. One of the first things people notice about you is how you carry and present yourself.

Politicians have always been ultra-aware of the importance and power of body language as a means of manipulating public perception. In the political world the *words* matter much less than the body language.

To highlight the way in which we are influenced (and therefore can influence) by a subliminal reading of a person's movements, researchers at Bangor University pioneered the use of the 'stick men' technique. They took a typical display of two politicians at the Despatch Box in the House of Commons – in this case Gordon Brown and David Cameron, the then prime minister and leader of the opposition, respectively. They then turned them into 'stick men' – so you had no idea of who the person was and in what capacity they were operating. Cameron's figure was propping itself up on one elbow while making sweeping gestures with the other arm. Brown was depicted with constant jabbing of the lower arm.

This was shown to people to judge which 'stick man' was more likeable and had better leadership qualities. David Cameron came across as the man who was more approachable, comfortable and relaxed. Gordon Brown came over as someone who was anxious, depressed, uncomfortable and aggressive.

So people take in a lot about you just from the way you move and gesticulate. Make sure you're sending the message you want to send.

You can see how a number of factors are at work in the 'first impression' stage and that we make decisions on another's likeability based on our *intuitive* feelings. We tend to make these evaluations instantly.

> I'm reminded of Andrew Lloyd Webber who – constantly plagued by journalists about why he wasn't liked more by the public – famously asked Alan J. Lerner, the *My Fair Lady* lyricist, if he could tell him why people took an 'instant' dislike to him.
> 'It saves time,' Lerner replied.

Body language during interviews

Interviews, especially job interviews, are terrifying for many people. In the case of an interview for employment, you're usually judged by a stranger who knows nothing about you and **observes your body language with little or no knowledge about your 'baseline' behaviour.**

How do you create a good impression by using 'good' body language? Well, you know what the opposite is because we've looked at – in great detail – all the negative 'leakage' gestures that can distort your message.

Try to ensure the following in order to make those all-important 'first impressions' work well for you.

- Remember that you're observed even *before* you meet the interviewer – your behaviour in the lift or in the reception area while you're waiting. So ensure that your behaviour is 'congruent' the whole time you're on show.
- Make sure your movements are with *purpose* as you enter the interviewer's office.
- Make sure the handshake grips the *palm* and not just the fingers.
- Adopt a *neutral smile* at first to show friendliness.
- Keep good *eye contact* at all times.
- Sometimes the *seating* causes problems. For example, a seat with a soft cushion or a sofa may push you down too low. Result – slumped posture, restricted speech and general air of submissiveness (not your fault). Try to get alternative seating, but otherwise sit forward on the edge and also lean forward (that shows attentiveness anyway).
- Listen with your *whole body*. Remember the appropriate head nods to show you're listening and to encourage the interviewer to continue when they are expressing something.
- Avoid *displacement* activities and *self-comfort* gestures.
- Just as your entrance was important, so is your exit.

As you approach the door make sure – unless you're after a '*does my bum look big in this*' assessment – that the final image of you is not your rear but your face. (Exceptions to this: interview at a bikini modelling agency, or for men an audition for 'Male Rear of the Year'.) As you make your exit, just turn around for your final goodbye.

Body language during a speech or presentation

If you have to give a speech or presentation, you'll know how important that 93 per cent (55 per cent visual and 38 per cent paralanguage) becomes.

- Try not to stand like a mannequin, especially if there's a podium. *Movement* keeps the audience awake.

- If there is a podium *try not to grip it tightly* as though you're on a 'big dipper' at a funfair – it conveys nervousness.

- Keep your head *up*.

- Arms out with the palms *up*.

- If you have to refer to a screen or flipchart don't forget to keep the *front* of the body *facing* the audience. Turn slightly and point.

- Try not to look *down* – either at notes or the floor.

- Make eye contact with all the audience *fleetingly* and *intermittently*. They want to feel as though you're speaking to them individually – it boosts your friendliness and credibility factor. (You know those radio presenters who make you feel as though they're talking just to you? Take a leaf out of their book.)

- Vary your 'paralinguistic' style by changing your rhythm throughout, with your pitch, tone and inflection.

- Look for any 'negative' body language in the audience by observing any *leakage* (now that you're an expert after the 7 Lessons!). Then act on it, either by involving the individual(s) or changing tack, or even scheduling a break if it's feasible.

Body language during flirting or dating

You can't even think about the words flirting or dating without mentioning body language. It reveals what the words from your mouth can't or won't say. Your body sends out the signals – some *deliberate*, some *subconscious* – to let the other person know you're interested, and then to get their attention and, after managing to do that, to keep them interested.

Women certainly have an advantage over men. Aside from generally being able to read body language better than men, they can use it to better effect. In the world of showbusiness you see it used to great effect. Screen legend Marilyn Monroe was applauded by some for her performance; others (mainly women!) criticised her physical brazenness. After her performance in the film that established her – in her famous pink 'wiggle' dress – *Niagara* (1953), actress Constance Bennett quipped: 'There's a broad with her future behind her.'

Marilyn Monroe

But Marilyn was well aware of the power of body language and the way she carried herself. She responded to her detractors: '*I walk the way I've always walked – since I was eleven or twelve.*' She provided further enlightenment for her critics: '*I'm a man's woman. I get along with women too. But I sashay up to a man; I walk up to a woman.*'

'Sashay'! It's fifty years in 2012 since the star's death. We haven't seen anyone like her since.

Men tend not to be as good as women at reading body language signals correctly. A display of friendliness may often be mistaken for sexual interest. A woman needs a man to interpret correctly the cues or signals she sends out (even if it's negative like 'shove off'), as does a man.

I'm discussing women first because all the research shows that it's *usually women who make the first move on a date or beginnings of a courtship.* **Men are usually reluctant to make contact with a woman who hasn't shown that she's approachable.**

As to the behaviour displayed by both sexes during the 'ritual', men try to behave in a more **manly** fashion and women try to appear more **feminine**.

Generally, if you get the 'big three' of approachability whereby a person *looks* directly at you, *smiles* and turns their body *towards* you, you can assume they're open to discussion. Most signals are more subtle, especially with women.

Her

Research shows the following activities (not foolproof, guys) to be common among women during flirting or courtship:

- A fleeting *smile* – a few times – in order to show they're approachable (*men often miss this signal*).
- Lowering or *dipping* of the head – makes the eyes seem bigger and gives a look of vulnerability.
- Wrist exposure – the delicate *underside of the wrist* is exposed frequently during the course of the conversation and the palm is exposed.
- The vulnerable neck area is *touched*.
- Knee is *pointed* in the direction of the person.
- Foot *pointed* in the person's direction (when standing).
- The *'leg twine'* (in the sitting position) – one leg pressed up against the other can emphasise sleekness of the legs.
- When catching the eye of men that they're potentially interested in, women may *flick their hair*.
- *Exposing* the neck by turning the head or lifting the chin.
- Pupils may *dilate* (involuntary).
- May *blink* more.
- Increase in *self-touching* – usually thigh, neck or throat.
- May toss her head back to flick hair over her shoulders or may run her *fingers through her hair*.
- Smoothing of *clothing*.

Interesting research here – the closer he is allowed to get to the handbag, the better the news!

Him

Men, by contrast, have very little in their repertoire compared to women. Research shows that men are not good at:

1 receiving signals
2 sending signals.

That doesn't leave much does it?

Men, mostly, just react to women's signals.

What are the activities common among men during the flirting and dating 'game' if one finds a woman attractive?

- *Posture* straightened.
- Stomach pulled *in* and chest *out*.
- Extended *gaze*.
- Gazes in 'flirting triangle' area of *mouth and below*.
- *Lifting* the head.
- May adopt a *deeper* voice at first.
- Hurried *preening* activity:
  - smoothing hair
  - straightening wristwatch
  - smoothing down clothes
  - fiddling with tie.

Good luck ladies!

Acting the part . . .

Actors on stage and screen are keenly aware of the power of body language in determining their performance. It's their posture, gestures and movements that ultimately determine their success. Take a tip from the experts. Make your body language say what you want to say.

Fifty years ago the two film producers Albert 'Cubby' Broccoli and Harry Saltzman were searching for an actor to play the part of secret agent 007, and interviewed a young hopeful called Sean Connery at their London headquarters in 1961. After a series of further interviews, despite reservations from other members of the team, Broccoli was convinced he had found their James Bond. He liked Connery's 'body language'.

A few years later in 1967 the two producers spoke at length in an interview about why they had chosen their (now superstar) man. They simply 'liked the way he moved'.

Saltzman went on to say that certain actors like Connery 'moved like cats . . . for a big man to be light on his feet is most unusual.'

In October 1962, *Dr No* opened in London and – from the opening scene in the casino – the film critics could see why the two producers realised they had found the man to lift the fictional hero from Ian Fleming's pages to the screen.

Eunice Gayson the first 'Bond girl' introduces herself ('*Trench...Silvia Trench*') to Sean Connery ('*Bond...James Bond*')

BODYtalk

Q So, this is what we've been leading up to. I never really thought about it. It looks as though it's body language that determines likeability in the final analysis.

I'm afraid so. That's why you're here. So ignore the things that we've discussed at your peril.

Q Yes, I can see I've got to be aware of voice now. I know how I subliminally 'turn off' mid-conversation if somebody's voice grates. I suppose I do speak with a bit of a monotone; probably sound even worse on the phone.

Yes, I'm sure.

Q Yes – and I can certainly see why girls have gone off to get another drink – **never to be seen on the planet again** – after I've been supposedly listening to them.

Suddenly everybody's talking about women. Is that why you're all here?

Q Well, in my case, partly. But surely everything we've discussed applies to normal dealings with people, at work, socially and – dare I say it – flirting and dating.

It does. Everything we've spoken about applies to your personal and professional life. In the case of romance, there are a few other 'gestures' and 'rules' that stray from normal social and work behaviour. Generally, men are not very good at giving out flirting signals – and not that great at picking them up. Too many ID 10T errors (and slapped faces).

Q In that case, can you give us men some signals to look out for? We'll assume the women know all about us.

I'll just throw a few at you. Obviously it's not exhaustive – we don't have that much time. But if you look for a cluster of activities, it may tell you something. Stop me if you want clarification.

- Exposing the wrist and palm.
- Head tilted to one side.
- Eyes dropping (during gaze) to the more intimate nose and mouth area, and even further down.
- Pupil size (dilation) – pupils will enlarge when we like what we see.
- Touching the neck or throat area.
- Pulling hair away to expose the neck and throat.
- Flicking hair back with a hand or with a head movement.
- Generally preening while maintaining eye contact with you.
- Leaning forward while standing and sitting.
- Feet pointing towards you.
- Adjusting and fidgeting with clothes.

Q So are these foolproof signs then?

Certainly not, if you go by just one gesture or activity. These are typically observed activities if a woman is interested (or may be interested) in a man. But – I'll have to repeat again – look for a few gestures to make up that 'sentence' we spoke about early on. One gesture is like a word – you can't derive meaning from it. A number of gestures – say three – can make a sentence, which then makes sense.

Q What about us women – don't we get a look in? Surely you emotionless men are capable of giving some signals when you're attracted to women?

Okay then ladies. We counter-intuitive, stony-faced men are not so expressive and there's not too much to go on. See if you can spot a cluster of these:

- An increase in self-grooming rituals (a few listed below).
- Eye gaze in the intimate nose and mouth area.
- Straightens or adjust a tie (or collar if there isn't one).
- Smoothes the hair.
- Pulls socks up.
- Fiddles with clothing – jacket button, shirt cuff, etc.
- Displays a lot of hand-to-face activity.

- Gives an eyebrow flash (fleeting).
- Generally displays a friendly and attentive face.

(Quite tame compared to the ladies, don't you think?)

Q Thanks for that. What about our working lives – I don't mean office romances, I'm off that subject now – I mean, how do we go about becoming likeable at work? Surely just being able to do the job well is enough.

Well, for a start you could work on all the body language features we've been discussing and also take heed of the research findings. The majority of employers and managers ultimately choose whether to employ someone or promote someone based, initially, on their likeability. Then, and only then, do they consider if the person is actually capable of doing the job.

Remember the body language rules!

1 Our mind-reading or 'thought-reading' derives from observing:
- eyes
- facial expressions
- gestures
- posture
- vocal characteristics.

2 Be aware of the 3 Cs at all times:
- context
- congruence
- clusters.

3 Warning: ID 10T error.

Observance of the 3 Cs is the key to your 'mind-reading'. Misinterpret and you'll be committing what's known (by me!) as the classic **ID 10T error.**

4 Remember the 55, 38, 7 rule.

5 Be aware of your own and other people's body language by looking for:
- displacement activities
- self-comfort gestures.

6 The key – when analysing body language – is always to look out for behaviour that indicates if a person is experiencing comfort or discomfort.

7 Every *thought* produces a *body* reaction.

Coffee break . . .

 Your likeability determines your success in both your personal and professional life.

 The secret of likeability is: how you make others feel.

 Research shows that if you make others feel good about themselves, through eye contact and paying attention, you are perceived as attractive.

 First impressions are important so it is no surprise that the first thing a person judges you on is your appearance.

 The smile comes out very high as a common denominator in the likeability stakes because it evokes a reciprocal response from the other person.

 Listening scores very highly in promoting likeability: women tend to be more animated when talking with each other and therefore tend to establish rapport quicker than men.

 Women tend to listen with the whole body, showing visual clues and promoting more empathy.

 Men tend to show fewer visual clues that give evidence of listening, so when a woman comes across a good male listener she's usually 'bowled over'!

 Like listening, good eye contact is a winner, but also expression in the eyes comes over as influential in the likeability equation.

Voice plays a big part too: if your pitch, tone and loudness are 'in sync' with their liking, you gain deeper rapport with the person.

 The biggest influence in your likeability 'quotient'? Why, your body language of course.

Afterword

We've come to the end of our 7 Lessons – but for all of you it's the beginning. The beginning of seeing things that were already out there, that you never perceived. You can now create the body language that is more effective for you by looking at what you did before and set about improving on it. **It's the small things that add up to make an overall impression.** Change those minor details and you'll find you've achieved major changes in terms of your effectiveness in dealing with people.

You can look at all the gestures and actions that we've spent time on and observe people in your everyday personal and working life, as they consciously and subconsciously bring body language to life. Except now, you know what every *body* is saying.

John Keats accused Isaac Newton of destroying all the poetry of the rainbow by explaining the origin of its colours, thereby dispelling its mystery. Some people believe that taking human behaviour and unravelling the mysteries of why we do things should be left alone. Body language is too important a subject for that to happen – I hope that with this updated knowledge you'll have better relationships. **You now know that you are truly able to read minds.** Remember:

- The mind produces a *thought*.
- The thought produces a *feeling*.
- That feeling 'leaks' out through the *body language*.
- You *read* the body language to ascertain a person's feelings. Hey presto, you're *mind-reading*.

Before signing off, just a reminder that when you've done everything you can, in the best possible way, of course, you will still come up against resistance from time to time. But it won't be because of your communication style. It will be the other person who is unable to 'read minds'. Your 'performance' will have been up to standard.

❝ Your 'performance' will have been up to standard ❞

You can't win them all, as the saying goes, but take heart from the message below:

'The play was a great success but the audience was a disaster.'

Oscar Wilde

Index

ALSO BY **JAMES BORG**

THE NUMBER ONE BESTSELLER
(Nielsen BookScan)

PERSUASION

Ever wondered what it's like to get people to do
whatever you want, whenever you want?

WINNER OF 'BEST OF THE REST' AWARD IN THE
800 CEO-READ BUSINESS BOOK AWARDS 2009 (USA)

SHORTLISTED FOR THE
BAA 'BEST NON-FICTION TRAVEL READ 2009 AWARD'

INTERNATIONAL BESTSELLER

PRAISE FOR *PERSUASION*

"One of the best-selling self-help/popular psychology books
of the 21st century"
Philip Stone, Charts Editor, *The Bookseller*

"This book is spot-on and should be a must read."
Daily Telegraph

"A rare 'self-help' book — marvellously readable and fun.
Hugely to be recommended."
Jilly Cooper

"This book should be on every individual's bookshelf."
Sir John Harvey-Jones

"An indispensable handbook for all of us who need to get other people to
do what we want."
Sir Anthony Jay, co-creator and writer of BBC's *Yes Minister*

"I'm persuaded that this book is an essential aid to getting people
on your side. Invaluable."
Sue Lawley, BBC radio and TV presenter

"A witty and fast-paced journey ... There are some real gems
in this book."
***Edge* Magazine**

"Gave me a new-found confidence ... won me £12000 of work ... helped
me gain more friends - all in 3 weeks. Give me more James Borg.
You are a bloody genius!"
Amazon reviewer

"Persuaded? We were. Buy it."
***Management Today* magazine (Voted 'best of its kind')**

"This is a handy readable guide ... The author persuaded
me to review this book. Damn, he is good."
Jeremy Vine, *The Times*

ALSO BY **JAMES BORG**

MIND POWER

Want to know how to change your life for the better
by changing the way you think?

"Completes his trilogy with a book packed with power".
Guardian

Our thoughts create our reality, so the quality of your thinking determines the quality of your life. Your mind really can propel you to success or hold you back. Your past, present and future is moulded by your thoughts.

Mind Power will show you how to free yourself from the thoughts that limit you in your personal and working life. You'll become acutely aware that when you change your thinking – you change your life.

"... *Persuasion* and *Body Language* have both been well-received and when you read *Mind Power* it's easy to see why ... intelligent and rational ways in which everyone can get their neurons firing and improve their thinking. Light-hearted and enthusiastic style make this one of the better self-help books out there."
Booksquawk

"*Mind Power* by James Borg is the best of the current self-help books."
Guardian